Summa Summarum

CMS Treatises in Mathematics

Published by the Canadian Mathematical Society

Traités de mathématiques de la SMC

Publié par la Société mathématique du Canada

Summa Summarum

Morgens Esrom Larsen
University of Copenhagen

Canadian Mathematical Society
Société mathématique du Canada
Ottawa, Ontario

A K Peters, Ltd.
Wellesley, Massachusetts

Sales and Customer Service

A K Peters, Ltd.
888 Worcester Street, Suite 230
Wellesley, MA 02482
www.akpeters.com

CMS Executive Office
Bureau administratif de la SMC

Canadian Mathematical Society
Société mathématique du Canada
577 King Edward
Ottawa, Ontario
Canada K1N 6N5
www.cms.math.ca/Publications

Library of Congress Cataloging-in-Publication Data

Larsen, Morgens Esrom.
Summa summarum / Morgens Esrom Larsen.
p. cm. -- (Treatises in mathematics)
Includes bibliographical references and index.
ISBN 13: 978-1-56881-323-3 (alk. paper)
ISBN 10: 1-56881-323-6 (alk. paper)
1. Exponential sums. 2. Combinatorial analysis. 3. Number theory. I. Title.

QA246.7.L37 2007
511'.6--dc22

2006103190

Printed in India
11 10 09 08 07

10 9 8 7 6 5 4 3 2 1

Dedicated to the memory of Erik Sparre Andersen (1919–2003)

Contents

Preface

"You never know when you will encounter a binomial coefficient sum."
—*Doron Zeilberger* [Zeilberger 91]

This textbook aims to provide a "summa", i.e., a collection, of all known algebraic finite sums and a guide to find the sum you need. Our hope is to find this summa on your desk—just as Thomas's original was found on the altar![1] Of course "all" is an exaggeration.

The kinds of sums we consider are often presented as sums of products of binomial coefficients, rational functions and occasionally harmonic numbers. Closed forms of such sums are usually called "combinatorial identities", though this term ought to include identities between sums, double sums and other equations too.

Our favorite approach is to recognize the identity in terms of some sum that is already known. To do so we need to write the sum in a standard form and to suggest a simple way to change a given sum to this standard form. The standard should be unique, hence we choose not to use more than one binomial coefficient, replacing any others with factorials. And the standard form should apply to the whole variety of sums considered, so we shall not use hypergeometric forms as they do not apply to the formulas of Chapters 14 and 15.

The ideal formula has the same form as the fundamental theorem of algebra, namely *a sum of products equals a product of sums.* Even so, we have to weaken this demand: often we can only attempt to write a sum of products as a sum with as few terms as possible of products of sums. The ideal is one term, but if such expressions simply fail to exist, sums of two or three similar terms may help.

If the desired sum is unknown, we aim to provide the reader with a variety of tools with which to attack the problem. Thus, our first few

[1]The title refers to Thomas Aquinas's *Summa Theologica.*

chapters introduce the basic tools of summation, and the remainder of the book provides a taxonomy for more general sums (Chapter 5) that includes an explanation of the famous algorithms of Gosper (Chapter 6) and Zeilberger (Chapter 14). To assist with the latter we include some guidance as to how to solve difference equations with rational coefficients. These later chapters are necessarily more concise, and at times the reader will have to do some work to fill in details.

In order to classify our summation formulas, we introduce a standard form which deviates from other recently popular forms. We dislike using hypergeometric forms for two reasons: they conceal several important properties, such as being symmetric, balanced or well-balanced, and they are insufficient for sums such as those due to Abel, Cauchy, Hagen–Rothe, and Jensen as well as sums containing harmonic numbers. We also dislike formulations which make extensive use of binomial coefficients, because they are not uniquely determined, and so it seems arbitrary to choose one form over another.

We prefer instead to use descending factorials, and have found it convenient to allow step lengths different from one. Hence we introduce the notation (see (1.2))

$$[x, d]_n = x(x-d)\cdots(x-(n-1)d),$$

which allows powers and ascending factorials to be written concisely as $[x, 0]_n$ and $[x, -1]_n$ respectively. In several cases we have found it natural to consider $d = 2$ as well. This is a true generalization, to be preferred to the analogy between powers and factorials as usually presented. We have also added an appendix on the most elementary generalizations to a q-basic form along with their proofs.

I am deeply indebted to my teacher in this field and dear friend, the late professor, Dr. Phil. Erik Sparre Andersen. Furthermore, I want to thank my colleague Professor Jørn Børling Olsson for improving my language. Lastly, I wish to thank my publishers for placing this book in an exciting new series.

Mogens Esrom Larsen
March 28, 2007

Chapter 1

Notation

1.1 Numbers

$\mathbb{N}$ = the natural numbers, $1, 2, 3, \ldots$.

$\mathbb{N}_0 = \mathbb{N} \cup \{0\}$, the numbers, $0, 1, 2, 3, \ldots$.

$\mathbb{Z}$ = the integers, $\ldots, -2, -1, 0, 1, 2, 3, \ldots$.

$\mathbb{R}$ = the real numbers.

$\mathbb{C}$ = the complex numbers, $z = x + iy$.

For the *nearest integers* to a real number, $x \in \mathbb{R}$, we use the standard notation for the *ceiling* and the *floor*,

$$\lceil x \rceil := \min\{n \in \mathbb{Z} \mid n \geq x\},$$
$$\lfloor x \rfloor := \max\{n \in \mathbb{Z} \mid n \leq x\},$$

and the *sign* $\sigma(x)$ of a real number, $x \in \mathbb{R}$, is defined by

$$\sigma(x) := \begin{cases} 1 & x \geq 0, \\ -1 & x < 0. \end{cases} \tag{1.1}$$

Furthermore, we denote the *maximum* and *minimum* of two numbers, $x, y \in \mathbb{R}$, as

$$x \vee y := \max\{x, y\},$$
$$x \wedge y := \min\{x, y\}.$$

1.2 Generalized Factorials

The *factorial* $[x, d]_n$ is defined for any *number* $x \in \mathbb{C}$, any *step size* $d \in \mathbb{C}$, and any *length* $n \in \mathbb{Z}$, except for $-x \in \{d, 2d, \ldots, -nd\}$, by

$$[x,d]_n := \begin{cases} \prod_{j=0}^{n-1} (x - jd) & n \in \mathbb{N}, \\ 1 & n = 0, \\ \prod_{j=1}^{-n} \dfrac{1}{x + jd} & -n \in \mathbb{N},\ -x \notin \{d, 2d, \ldots, -nd\}. \end{cases} \tag{1.2}$$

As special cases, we remark that

$$[x, 0]_n = x^n \quad n \in \mathbb{Z},$$

and furthermore, we want to apply the shorthands

$$[x]_n := [x, 1]_n \qquad \big(x(x-1)\cdots(x-n+1) \text{ for } n > 0\big), \tag{1.3}$$
$$(x)_n := [x, -1]_n \qquad \big(x(x+1)\cdots(x+n-1) \text{ for } n > 0\big). \tag{1.4}$$

The *binomial coefficients* $\binom{x}{n}$ are defined for $x \in \mathbb{C}$ and $n \in \mathbb{Z}$ by

$$\binom{x}{n} := \begin{cases} \dfrac{[x]_n}{[n]_n} & \text{for } n \in \mathbb{N}_0, \\ 0 & \text{for } -n \in \mathbb{N}. \end{cases} \tag{1.5}$$

1.3 Differences

The *identity operator* is defined for $f\colon \mathbb{Z} \to \mathbb{C}$ as

$$\mathbf{I}(f)(k) := f(k),$$

and the *shift operator* is defined as

$$\mathbf{E}(f)(k) := f(k+1). \tag{1.6}$$

From these we define the *difference operator* Δ by the expressions

$$\Delta := \mathbf{E} - \mathbf{I}, \quad \Delta f(k) := f(k+1) - f(k). \tag{1.7}$$

An *indefinite sum* or *anti-difference* of a function $g(k)$ is defined as any solution $f(k)$ to the equation

$$g(k) = \Delta f(k).$$

(The anti-difference $f(k)$ is uniquely determined up to a constant, or a periodic function with period 1, by the function $g(k)$.)

We denote this indefinite sum by

$$\sum g(k)\delta k := f(k). \tag{1.8}$$

We use δx in analogy to the standard dx notation for integrals.

For any indefinite sum (1.8) of a function $g(k)$, a *definite sum* is defined to be

$$\sum\nolimits_a^b g(k)\delta k := f(b) - f(a), \tag{1.9}$$

and we remark that the connection to the usual step-by-step sum is

$$\sum\nolimits_a^b g(k)\delta k = \sum_{k=a}^{b-1} g(k). \tag{1.10}$$

The *harmonic* numbers are defined by

$$H_n := \sum_{k=1}^{n} \frac{1}{k} = \sum\nolimits_0^n [k]_{-1}\delta k. \tag{1.11}$$

The *generalized harmonic* numbers are defined for $n, m \in \mathbb{N}$ and $c \in \mathbb{C}$ by

$$H_{c,n}^{(m)} := \sum_{k=1}^{n} \frac{1}{(c+k)^m}. \tag{1.12}$$

Note that $H_{0,n}^{(1)} = H_n$.

Chapter 2

Elementary Properties

2.1 Factorial Formulas

The factorial satisfies some obvious, but very useful, rules of computation. The most important ones are

$$[x,d]_k = [-x+(k-1)d,d]_k(-1)^k \qquad x,d\in\mathbb{C},\ k\in\mathbb{Z}, \tag{2.1}$$

$$[x,d]_k = [x,d]_h[x-hd,d]_{k-h} \qquad x,d\in\mathbb{C},\ k,h\in\mathbb{Z}, \tag{2.2}$$

$$[x,d]_k = 1/[x-kd,d]_{-k} \qquad x,d\in\mathbb{C},\ k\in\mathbb{Z}, \tag{2.3}$$

$$[x,d]_k = [x-d,d]_k + kd[x-d,d]_{k-1} \qquad x,d\in\mathbb{C},\ k\in\mathbb{Z}, \tag{2.4}$$

$$[xd,d]_k = d^k[x]_k \qquad x,d\in\mathbb{C},\ k\in\mathbb{Z}. \tag{2.5}$$

Applying the difference operator (1.7), we can rewrite (2.4) for $d=1$ as

$$\Delta[k]_n = n[k]_{n-1}.$$

For sums, this formula gives us, by (1.9),

$$\sum\nolimits_0^m [k]_n\,\delta k = \begin{cases} \dfrac{[m]_{n+1}}{n+1} - \dfrac{[0]_{n+1}}{n+1} & \text{for } n\neq -1, \\ H_m & \text{for } n=-1, \end{cases} \tag{2.6}$$

using the harmonic numbers (1.11).

2.2 Binomial Coefficient Formulas

Similarly, the binomial coefficients (1.5) satisfy a series of rules:

$$\binom{x}{k} = \binom{x-1}{k-1} + \binom{x-1}{k} \qquad x \in \mathbb{C},\ k \in \mathbb{N}_0, \tag{2.7}$$

$$\binom{x}{k}[k]_m = [x]_m\binom{x-m}{k-m} \qquad x \in \mathbb{C},\ m \in \mathbb{Z},\ k \in \mathbb{N}_0, \tag{2.8}$$

$$\binom{x}{k}\binom{k}{m} = \binom{x}{m}\binom{x-m}{k-m} \qquad x \in \mathbb{C},\ k, m \in \mathbb{Z}, \tag{2.9}$$

$$\binom{x}{k}[y]_k = [x]_k\binom{y}{k} \qquad x, y \in \mathbb{C},\ k \in \mathbb{Z}, \tag{2.10}$$

$$\binom{x}{k} = (-1)^k\binom{k-x-1}{k} \qquad x \in \mathbb{C},\ k \in \mathbb{Z}, \tag{2.11}$$

$$\binom{m}{k} = \binom{m}{m-k} \qquad m \in \mathbb{N}_0,\ k \in \mathbb{Z}. \tag{2.12}$$

2.3 Anti-Differences

Applying the difference operator (1.7), we can rewrite (2.7) as

$$\Delta\binom{k}{n} = \binom{k}{n-1},$$

equivalent to

$$\sum\binom{k}{n}\delta k = \binom{k}{n+1}. \tag{2.13}$$

If we introduce the alternating sign, (2.7) can be written

$$\Delta(-1)^k\binom{x}{k} = (-1)^{k+1}\binom{x+1}{k+1}, \tag{2.14}$$

which is equivalent to

$$\sum(-1)^k\binom{x}{k}\delta k = (-1)^{k+1}\binom{x-1}{k-1}.$$

We omit the proofs, since the formulas (2.7)–(2.12) are easily proved using (1.5) and the elementary properties (2.1)–(2.5) of the factorials.

The binomial coefficient itself has a "less nice" difference, but we note that

$$\Delta\binom{x-1}{k-1} = \binom{x}{k}\left(1 - 2\tfrac{k}{x}\right).$$

Indeed, by (2.4) we have

$$\begin{aligned}\Delta\binom{x-1}{k-1} &= \binom{x-1}{k} - \binom{x-1}{k-1} = \frac{[x-1]_k - k[x-1]_{k-1}}{[k]_k}\\ &= \frac{[x-1]_{k-1}}{[k]_k}\,(x-1-k+1-k) = \frac{[x]_k}{[k]_k}\frac{1}{x}\,(x-2k)\\ &= \binom{x}{k}\left(1-2\tfrac{k}{x}\right).\end{aligned}$$

The difference and shift operators in (1.7) and (1.6) commute and satisfy

$$\Delta(fg) = f\Delta g + \mathbf{E}g\Delta f \tag{2.15}$$

and

$$\begin{aligned}\mathbf{E}k - k\mathbf{E} &= \mathbf{E},\\ \Delta k - k\Delta &= \mathbf{E},\end{aligned}$$

where k is the operator $(k \to f(k)) \to (k \to kf(k))$. This just means that $(k+1)f(k+1) - kf(k+1) = f(k+1)$, etc.

The definite sum (1.9) is

$$\sum_{k=0}^{n} g(k) = \sum\nolimits_{0}^{n+1} g(k)\,\delta k.$$

Summation by parts is given by (2.15) as

$$\sum f(k)\Delta g(k)\,\delta k = f(k)g(k) - \sum \mathbf{E}g(k)\Delta f(k)\,\delta k, \tag{2.16}$$

$$\sum\nolimits_a^b f(k)\Delta g(k)\,\delta k = f(b)g(b) - f(a)g(a) - \sum\nolimits_a^b g(k+1)\Delta f(k)\,\delta k, \tag{2.17}$$

and is also called *Abelian summation.*

As we have the identities

$$\begin{aligned}\Delta &= \mathbf{E} - \mathbf{I},\\ \mathbf{E} &= \Delta + \mathbf{I},\end{aligned}$$

for $n \in \mathbb{N}$, we get the obvious relations between their iterations:

$$\Delta^n = \sum_{k=0}^{n} \binom{n}{k} (-1)^k \mathbf{E}^{n-k},$$

$$\mathbf{E}^n = \sum_{k=0}^{n} \binom{n}{k} \Delta^k.$$

The following will be referred to as *inversion*:[1]

Given a sequence $f(k)$, $k \in \mathbb{N}_0$, *let*

$$g(n) = \sum_{k=0}^{n} (-1)^k \binom{n}{k} f(k). \tag{2.18}$$

Then

$$f(n) = \sum_{k=0}^{n} (-1)^k \binom{n}{k} g(k). \tag{2.19}$$

Proof: Substituting (2.18) into (2.19) and using (2.9),

$$\begin{aligned}
\sum_{k=0}^{n} (-1)^k \binom{n}{k} g(k) &= \sum_{k=0}^{n} (-1)^k \binom{n}{k} \sum_{j=0}^{k} (-1)^j \binom{k}{j} f(j) \\
&= \sum_{j=0}^{n} \sum_{k=j}^{n} (-1)^k \binom{n}{k} (-1)^j \binom{k}{j} f(j) \\
&= \sum_{j=0}^{n} f(j) \binom{n}{j} \sum_{k=j}^{n} (-1)^{k-j} \binom{n-j}{k-j} \\
&= \sum_{j=0}^{n} f(j) \binom{n}{j} 0^{n-j} = f(n).
\end{aligned}$$

□

[1] This is a special case of *Möbius inversion* in the Boolean algebra $\mathbf{2}^n$ of all subsets of an n-element set.

Chapter 3

Polynomials

3.1 Geometric Progression

For $q \neq 1$ we have

$$\sum_{k=0}^{n} q^k = \frac{1 - q^{n+1}}{1 - q}, \tag{3.1}$$

or more generally

$$\sum_{k=m}^{n} q^k = q^m \frac{1 - q^{n-m+1}}{1 - q}. \tag{3.2}$$

Of course, for $q = 1$, we get

$$\sum_{k=0}^{n} q^k = \sum_{k=0}^{n} 1 = n + 1.$$

Proof: We have $\Delta q^k = q^{k+1} - q^k = (q-1)q^k$. So according to (1.10) and (1.9) we get (3.2):

$$\sum_{k=m}^{n} q^k = \sum\nolimits_m^{n+1} q^k \delta k = \frac{q^{n+1}}{q-1} - \frac{q^m}{q-1} = \frac{q^{n+1} - q^m}{q-1} = q^m \frac{1 - q^{n-m+1}}{1 - q}.$$

□

3.2 Sums of Polynomials

According to an anecdote, C. F. Gauss (1777–1855) went to school at the age of seven. One day the teacher, in order to keep the class occupied, asked the students to add up all the numbers from one to one hundred. Gauss immediately answered 5050. He wrote the numbers in two lines:

$$\begin{array}{ccccccccc} 1 & + & 2 & + & 3 & + & \cdots & + & 100, \\ 100 & + & 99 & + & 98 & + & \cdots & + & 1. \end{array}$$

Adding the pairs, he obtained the sum 101, for 100 pairs. Thus he obtained 100100, which is twice the desired sum. It could be written like this:

$$\sum_{k=1}^{100} k = \sum_{k=1}^{100}(101-k) = \frac{1}{2}\sum_{k=1}^{100}(k+101-k) = \frac{1}{2}\sum_{k=1}^{100} 101 = \frac{1}{2}100\times 101 = 5050.$$

We have learned something: the advantage of changing the order of summation, and the formula

$$\sum_{k=1}^{n} k = \frac{n(n+1)}{2}. \tag{3.3}$$

If we have a polynomial $p(k) = a_0 + a_1 k + \cdots + a_m k^m$ and want to determine $\sum_{k=0}^{n} p(k)$, we need to know the sums $\sum_{k=0}^{n} k^m$ for all $m \geq 0$. We found the formula for $m = 1$ above. The formulas for small values of m were determined in 1631 by J. Faulhaber (1580–1635) [Knuth 93]. In the following, N is an abbreviation for $n(n+1)$.

$$\sum_{k=1}^{n} k^0 = n, \tag{3.4}$$

$$\sum_{k=1}^{n} k^1 = \frac{N}{2}, \tag{3.5}$$

$$\sum_{k=1}^{n} k^2 = \frac{N(2n+1)}{2\cdot 3}, \tag{3.6}$$

$$\sum_{k=1}^{n} k^3 = \frac{N^2}{4}, \tag{3.7}$$

$$\sum_{k=1}^{n} k^4 = \frac{N p_4(N)(2n+1)}{2\cdot 3\cdot 5}, \quad p_4(x) = 3x - 1, \tag{3.8}$$

$$\sum_{k=1}^{n} k^5 = \frac{N^2 p_5(N)}{2\cdot 6}, \quad p_5(x) = 2x - 1, \tag{3.9}$$

$$\sum_{k=1}^{n} k^6 = \frac{Np_6(N)(2n+1)}{2\cdot 3\cdot 7}, \quad p_6(x) = 3x^2 - 3x + 1, \tag{3.10}$$

$$\sum_{k=1}^{n} k^7 = \frac{N^2p_7(N)}{3\cdot 8}, \quad p_7(x) = 3x^2 - 4x + 2, \tag{3.11}$$

$$\sum_{k=1}^{n} k^8 = \frac{Np_8(N)(2n+1)}{2\cdot 5\cdot 9}, \quad p_8(x) = 5x^3 - 10x^2 + 9x - 3, \tag{3.12}$$

$$\sum_{k=1}^{n} k^9 = \frac{N^2p_9(N)}{2\cdot 10}, \quad p_9(x) = 2x^3 - 5x^2 + 6x - 3, \tag{3.13}$$

$$\sum_{k=1}^{n} k^{10} = \frac{Np_{10}(N)(2n+1)}{2\cdot 3\cdot 11}, \quad p_{10}(x) = 3x^4 - 10x^3 + 17x^2 - 15x + 5, \tag{3.14}$$

$$\sum_{k=1}^{n} k^{11} = \frac{N^2p_{11}(N)}{2\cdot 12}, \quad p_{11}(x) = 2x^4 - 8x^3 + 17x^2 - 20x + 10, \tag{3.15}$$

$$\sum_{k=1}^{n} k^{12} = \frac{Np_{12}(N)(2n+1)}{2\cdot 3\cdot 5\cdot 7\cdot 13}, \quad p_{12}(x) = 105x^5 - 525x^4 + 1435x^3 - 2360x^2 + 2073x - 691, \tag{3.16}$$

$$\sum_{k=1}^{n} k^{13} = \frac{N^2p_{13}(N)}{2\cdot 3\cdot 5\cdot 14}, \quad p_{13}(x) = 30x^5 - 175x^4 + 574x^3 - 1180x^2 + 1382x - 691, \tag{3.17}$$

$$\sum_{k=1}^{n} k^{14} = \frac{Np_{14}(N)(2n+1)}{2\cdot 3\cdot 15}, \quad p_{14}(x) = 3x^6 - 21x^5 + 84x^4 - 220x^3 + 359x^2 - 315x + 105, \tag{3.18}$$

$$\sum_{k=1}^{n} k^{15} = \frac{N^2p_{15}(N)}{3\cdot 16}, \quad p_{15}(x) = 3x^6 - 24x^5 + 112x^4 - 352x^3 + 718x^2 - 840x + 420, \tag{3.19}$$

$$\sum_{k=1}^{n} k^{16} = \frac{Np_{16}(N)(2n+1)}{2\cdot 3\cdot 5\cdot 17}, \quad p_{16}(x) = 15x^7 - 140x^6 + 770x^5 - 2930x^4 + 7595x^3 - 12370x^2 + 10851x - 3617, \tag{3.20}$$

$$\sum_{k=1}^{n} k^{17} = \frac{N^2p_{17}(N)}{2\cdot 5\cdot 18}, \quad p_{17}(x) = 10x^7 - 105x^6 + 660x^5 - 2930x^4 + 9114x^3 - 18555x^2 + 21702x - 10851, \tag{3.21}$$

$$\sum_{k=1}^{n} k^{18} = \frac{Np_{18}(N)(2n+1)}{2\cdot 3\cdot 5\cdot 7\cdot 19}, \quad \begin{aligned} p_{18}(x) = 105x^8 &-1260x^7+9114x^6 \\ &-47418x^5+178227x^4-460810x^3 \\ &+750167x^2-658005x+219335, \end{aligned} \tag{3.22}$$

$$\sum_{k=1}^{n} k^{19} = \frac{N^2p_{19}(N)}{2\cdot 3\cdot 7\cdot 20}, \quad \begin{aligned} p_{19}(x) = 42x^8 &-560x^7+4557x^6 \\ &-27096x^5+118818x^4-368648x^3 \\ &+750167x^2-877340x+438670. \end{aligned} \tag{3.23}$$

In principle, the proofs of all these formulas are trivial. For example, let us take (3.6):

$$\sum_{k=1}^{n} k^2 = \sum_{k=0}^{n+1} k^2 \delta k = \frac{N(2n+1)}{2\cdot 3} = \frac{n(n+1)(2n+1)}{2\cdot 3}.$$

Taking the difference of the right side, we obtain

$$\begin{aligned} \Delta\frac{(k-1)k(2k-1)}{2\cdot 3} &= \frac{k(k+1)(2k+1)}{2\cdot 3} - \frac{k(k-1)(2k-1)}{2\cdot 3} \\ &= \frac{k(2k^2+3k+1-2k^2+3k-1)}{6} \\ &= k^2. \end{aligned}$$

3.3 Sums of Polynomials by Bernoulli Polynomials

The Faulhaber formulas are special cases of a much more powerful discovery of J. Bernoulli (1654–1705), see *Ars Conjectandi* [Bernoulli 13].

The Bernoulli polynomials are polynomial solutions to the equations

$$\Delta f_n(k) = nk^{n-1}, \quad n \in \mathbb{N}, \tag{3.24}$$

and are uniquely determined up to the constant term. For any solution to (3.24), from (1.8) we then get formulas for the sums

$$\sum k^n \delta k = \frac{B_{n+1}(k)}{n+1}.$$

The solution goes as follows. Differentiation of (3.24) with respect to k yields

$$\Delta f_n'(k) = n(n-1)k^{n-2},$$

proving that $\frac{1}{n}f_n'(k)$ solves (3.24) for $n-1$, hence that

$$\frac{1}{n}f_n'(k)-f_{n-1}(k) \tag{3.25}$$

is a constant.

The polynomial solutions to (3.24) for which the constant term is zero are called the *Bernoulli polynomials* and are denoted by $B_n(k)$.

Suppose we have written the Bernoulli polynomials in the form

$$B_n(k) = \sum_{j=0}^{n} \binom{n}{j} \beta_{n-j}^n k^j \tag{3.26}$$

for suitable constants β_{n-j}^n. The index is chosen so that β_0^n is the coefficient of the leading term k^n, and β_n^n is the constant term.

Differentiation of B_n yields

$$\begin{aligned} B_n'(k) &= \sum_{j=1}^{n} \binom{n}{j} \beta_{n-j}^n j k^{j-1} \\ &= \sum_{j=1}^{n} n \binom{n-1}{j-1} \beta_{n-j}^n k^{j-1} \\ &= n \sum_{j=0}^{n-1} \binom{n-1}{j} \beta_{n-1-j}^n k^j. \end{aligned}$$

From (3.25), this is known to be equal to

$$nB_{n-1}(k) = n \sum_{j=0}^{n-1} \binom{n-1}{j} \beta_{n-1-j}^{n-1} k^j.$$

The conclusion from comparing the coefficients is that

$$\beta_{n-1-j}^n = \beta_{n-1-j}^{n-1}, \qquad \text{for } j = 0, 1, \ldots, n-1. \tag{3.27}$$

The values in (3.27) are called the *Bernoulli numbers* and are denoted by B_{n-1-j}, omitting the now superfluous superscript.

Rewriting (3.26) in terms of the Bernoulli numbers gives

$$B_n(k) = \sum_{j=0}^{n} \binom{n}{j} B_{n-j} k^j,$$

The fact that the Bernoulli polynomials satisfy (3.24) will lead to a computation of their coefficients. We compute

$$
\begin{aligned}
\Delta B_n(k) &= \sum_{j=0}^{n}\binom{n}{j}B_{n-j}((k+1)^j - k^j) = \sum_{j=0}^{n}\binom{n}{j}B_{n-j}\sum_{i=0}^{j-1}\binom{j}{i}k^i \\
&= \sum_{i=0}^{n-1}k^i\sum_{j=i+1}^{n}\binom{n}{j}\binom{j}{i}B_{n-j} = \sum_{i=0}^{n-1}k^i\sum_{j=i+1}^{n}\binom{n}{i}\binom{n-i}{j-i}B_{n-j} \\
&= \sum_{i=0}^{n-1}\binom{n}{i}k^i\sum_{j=i+1}^{n}\binom{n-i}{j-i}B_{n-j} \\
&= \sum_{i=0}^{n-1}\binom{n}{n-i}k^i\sum_{j=1}^{n-i}\binom{n-i}{j}B_{n-i-j} = \sum_{\ell=1}^{n}\binom{n}{\ell}k^{n-\ell}\sum_{j=1}^{\ell}\binom{\ell}{j}B_{\ell-j} \\
&= \sum_{\ell=1}^{n}\binom{n}{\ell}k^{n-\ell}\sum_{i=0}^{\ell-1}\binom{\ell}{i}B_i,
\end{aligned}
$$

where we have applied the binomial formula (7.1) and formulas (2.9), (2.12), and have reversed the direction of summation, e.g., $\ell = n-i$ and $i = \ell-j$.

Now we know from (3.24) that this final polynomial equals nk^{n-1}. This means that the coefficients are n for $\ell = 1$ and 0 otherwise. This gives the formulas for the Bernoulli numbers:

$$B_0 = 1, \quad \sum_{i=0}^{\ell-1}\binom{\ell}{i}B_i = 0 \text{ for } \ell > 1. \tag{3.28}$$

Some authors add the number B_ℓ to the last sum to get the "implicit" recursion formula

$$\sum_{i=0}^{\ell}\binom{\ell}{i}B_i = B_\ell. \tag{3.29}$$

Perhaps it looks nicer, but it can confuse the reader.

Either formula (3.28) or (3.29) allows the computation of the Bernoulli numbers; for the first few (noting that $B_i = 0$ for odd $i \geq 3$), we get

$$B_0 = 1, \quad B_1 = -\frac{1}{2}, \quad B_2 = \frac{1}{6}, \quad B_4 = -\frac{1}{30}, \quad B_6 = \frac{1}{42},$$

$$B_8 = -\frac{1}{30}, \quad B_{10} = \frac{5}{66}, \quad B_{12} = -\frac{691}{2730}.$$

As soon as we have the numbers, we can write down the polynomials.

$$
\begin{aligned}
B_0(k) &= 1,\\
B_1(k) &= k-\tfrac{1}{2},\\
B_2(k) &= k^2-k+\tfrac{1}{6},\\
B_3(k) &= k(k-1)\left(k-\tfrac{1}{2}\right),\\
B_4(k) &= k^4-2k^3+k^2-\tfrac{1}{30},\\
B_5(k) &= k(k-1)\left(k-\tfrac{1}{2}\right)\left(k^2-k-\tfrac{1}{3}\right),\\
B_6(k) &= k^6-3k^5+\tfrac{5}{2}k^3-\tfrac{1}{2}k^2+\tfrac{1}{42},\\
B_7(k) &= k(k-1)\left(k-\tfrac{1}{2}\right)\left(k^4-2k^3+2k+\tfrac{1}{3}\right),\\
B_8(k) &= k^8-4k^7+\tfrac{14}{3}k^6-\tfrac{7}{3}k^4+\tfrac{2}{3}k^2-\tfrac{1}{30},\\
B_9(k) &= k(k-1)\left(k-\tfrac{1}{2}\right)\left(k^6-3k^5+k^4-\tfrac{1}{5}k^2-\tfrac{9}{5}k-\tfrac{3}{5}\right),\\
B_{10}(k) &= k^{10}-5k^9+\tfrac{15}{2}k^8-7k^6+5k^4-\tfrac{3}{2}k^2+\tfrac{5}{66}.
\end{aligned}
$$

Figure 3.1 shows graphs of the first four Bernoulli polynomials. In fact, all odd Bernoulli polynomials look like $\pm B_3$, and the even ones look like $\pm B_4$, i.e., the odd ones for $n \geq 3$ have 3 zeros in the interval $[0,1]$ and the even ones have 2 zeros. But the Bernoulli numbers grow quickly in absolute value, e.g.,

$$
B_{20} = -\frac{174611}{330}, \quad B_{40} = -\frac{261082718496449122051}{13530},
$$
$$
B_{60} = -\frac{1215233140483755572040304994079820246041491}{56786730}.
$$

In fact, they grow like

$$
\frac{2(2n)!}{(2\pi)^{2n}}.
$$

The fact that

$$
B_n'(k) = nB_{n-1}(k)
$$

yields the general formula for the derivatives of the Bernoulli polynomials:

$$
B_n^{(j)}(k) = [n]_j B_{n-j}(k).
$$

This formula provides the Taylor development of the polynomials as

$$
B_n(k+h) = \sum_{j=0}^{n} [n]_j B_{n-j}(k) \frac{h^j}{j!} = \sum_{j=0}^{n} \binom{n}{j} B_{n-j}(k) h^j.
$$

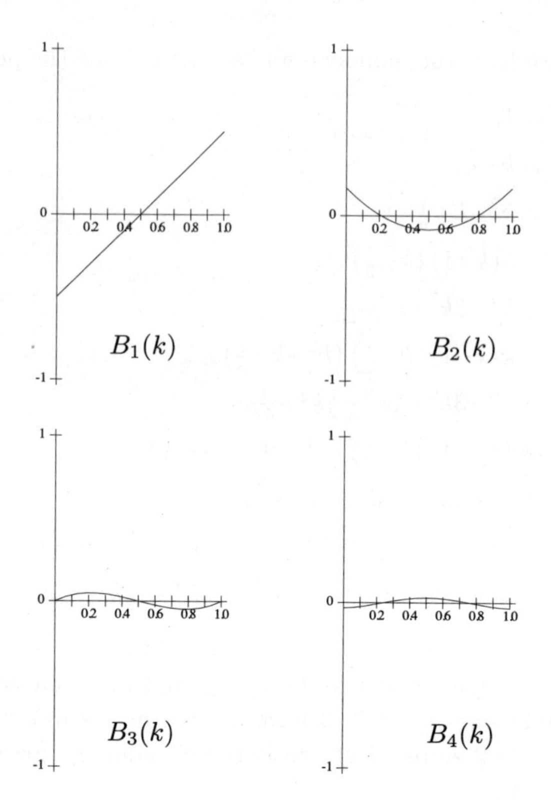

Figure 3.1. Bernoulli polynomials B_1, B_2, B_3, B_4

With the choice of $h = 1$, we get from the defining equation (3.24) that

$$nk^{n-1} = B_n(k+1) - B_n(k) = \sum_{j=1}^{n} \binom{n}{j} B_{n-j}(k). \tag{3.30}$$

Consider the function

$$f(k) = (-1)^{n+1} B_{n+1}(1-k).$$

We get immediately

$$\begin{aligned} f(k+1) - f(k) &= (-1)^{n+1} \left(B_{n+1}(-k) - B_{n+1}(1-k) \right) \\ &= -(-1)^{n+1}(n+1)(-k)^n = (n+1)k^n. \end{aligned}$$

This means that $f(k)$ deviates from $B_{n+1}(k)$ by a constant. Rather than finding this constant, we just differentiate the two functions and obtain the formula

$$B_n(1-k) = (-1)^n B_n(k).$$

The equation

$$f(k+\tfrac{1}{2}) - f(k) = (n+1)k^n$$

has the two solutions $B_{n+1}(k)+B_{n+1}(k+\frac{1}{2})$ and $2^{-n}B_{n+1}(2k)$, both polynomials in k. Their difference is again a polynomial which is periodic and hence constant. So differentiation of these two solutions yields the same function, and hence we get the formula

$$B_n(2k) = 2^{n-1}\left(B_n(k) + B_n(k+\tfrac{1}{2})\right).$$

For $k=0$, we get the value

$$B_n(\tfrac{1}{2}) = -\left(1-2^{1-n}\right)B_n.$$

For n even, the absolute value at $\frac{1}{2}$ is just a little smaller than the absolute value at 0 and 1.

3.4 Sums of Polynomials by Stirling Numbers

The sums of a polynomial can be found by writing the polynomial in k in terms of $[k]_m$, $m = 0, 1, \ldots$. This can be accomplished by the use of the *Stirling numbers of the second kind,* after J. Stirling (1692–1770):

$$\begin{pmatrix} k \\ k^2 \\ k^3 \\ k^4 \\ k^5 \\ k^6 \\ k^7 \\ k^8 \end{pmatrix} = \begin{pmatrix} 1 & 0 & 0 & 0 & 0 & 0 & 0 & 0 \\ 1 & 1 & 0 & 0 & 0 & 0 & 0 & 0 \\ 1 & 3 & 1 & 0 & 0 & 0 & 0 & 0 \\ 1 & 7 & 6 & 1 & 0 & 0 & 0 & 0 \\ 1 & 15 & 25 & 10 & 1 & 0 & 0 & 0 \\ 1 & 31 & 90 & 65 & 15 & 1 & 0 & 0 \\ 1 & 63 & 301 & 350 & 140 & 21 & 1 & 0 \\ 1 & 127 & 966 & 1701 & 1050 & 266 & 28 & 1 \end{pmatrix} \cdot \begin{pmatrix} [k]_1 \\ [k]_2 \\ [k]_3 \\ [k]_4 \\ [k]_5 \\ [k]_6 \\ [k]_7 \\ [k]_8 \end{pmatrix}. \tag{3.31}$$

This matrix can be extended indefinitely, see [Abramowitz and Stegun 65, p. 835]. It simply says that, for instance,

$$k^5 = 1\times[k]_1 + 15\times[k]_2 + 25\times[k]_3 + 10\times[k]_4 + 1\times[k]_5.$$

The Stirling numbers of the second kind are denoted by $\mathfrak{S}_n^{(j)}$ and appear as coefficients in the formula

$$k^n = \sum_{j=1}^{n} \mathfrak{S}_n^{(j)}[k]_j. \tag{3.32}$$

From (3.32), multiplying by k we get

$$\begin{aligned} k^{n+1} &= \sum_{j=1}^{n} \mathfrak{S}_n^{(j)} k[k]_j = \sum_{j=1}^{n} \mathfrak{S}_n^{(j)} (k-j+j)[k]_j \\ &= \sum_{j=1}^{n} \mathfrak{S}_n^{(j)} [k]_{j+1} + \sum_{j=1}^{n} \mathfrak{S}_n^{(j)} j[k]_j = \sum_{j=1}^{n+1} \mathfrak{S}_n^{(j-1)} [k]_j + \sum_{j=1}^{n} \mathfrak{S}_n^{(j)} j[k]_j \\ &= \sum_{j=1}^{n+1} \left(\mathfrak{S}_n^{(j-1)} + j\mathfrak{S}_n^{(j)} \right) [k]_j . \end{aligned}$$

From this we derive the recurrence formula

$$\mathfrak{S}_{n+1}^{(j)} = \mathfrak{S}_n^{(j-1)} + j\mathfrak{S}_n^{(j)} .$$

This formula shows that the Stirling numbers of the second kind can be interpreted as follows: $\mathfrak{S}_n^{(j)}$ *is the number of ways a set of n objects can be divided into j nonempty subsets.*

It can also be used to prove the explicit formula,

$$\mathfrak{S}_n^{(j)} = \frac{1}{j!} \sum_{k=0}^{n} (-1)^{j-k} \binom{j}{k} k^n .$$

The Stirling numbers of the first kind are the solutions to the inverse problem, i.e., the coefficients of the expressions of the factorials $[k]_n$ in terms of the monomials, i.e.,

$$[k]_n = \sum_{j=1}^{n} S_n^{(j)} k^j . \tag{3.33}$$

They can also be conveniently arranged in a matrix:

$$\begin{pmatrix} [k]_1 \\ [k]_2 \\ [k]_3 \\ [k]_4 \\ [k]_5 \\ [k]_6 \\ [k]_7 \\ [k]_8 \end{pmatrix} = \begin{pmatrix} 1 & 0 & 0 & 0 & 0 & 0 & 0 & 0 \\ -1 & 1 & 0 & 0 & 0 & 0 & 0 & 0 \\ 2 & -3 & 1 & 0 & 0 & 0 & 0 & 0 \\ -6 & 11 & -6 & 1 & 0 & 0 & 0 & 0 \\ 24 & -50 & 35 & -10 & 1 & 0 & 0 & 0 \\ -120 & 274 & -225 & 85 & -15 & 1 & 0 & 0 \\ 720 & -1764 & 1624 & -735 & 175 & -21 & 1 & 0 \\ -5040 & 13068 & -13132 & 6769 & -1960 & 322 & -28 & 1 \end{pmatrix} \cdot \begin{pmatrix} k \\ k^2 \\ k^3 \\ k^4 \\ k^5 \\ k^6 \\ k^7 \\ k^8 \end{pmatrix} .$$

This just states that for example,

$$[k]_5 = k^5 - 10k^4 + 35k^3 - 50k^2 + 24k.$$

The matrix is simply the inverse of the matrix in (3.31). The Stirling numbers can be found in [Abramowitz and Stegun 65, p. 833].

If we multiply (3.33) by $k-n$, we obtain

$$\begin{aligned}[k]_{n+1} &= \sum_{j=1}^{n} S_n^{(j)}(k-n)k^j = \sum_{j=1}^{n} S_n^{(j)}k^{j+1} - \sum_{j=1}^{n} S_n^{(j)}nk^j \\ &= \sum_{j=1}^{n+1}\left(S_n^{(j-1)} - nS_n^{(j)}\right)k^j.\end{aligned}$$

From this we derive the recurrence formula

$$S_{n+1}^{(j)} = S_n^{(j-1)} - nS_n^{(j)}. \tag{3.34}$$

This formula shows that the absolute value of the Stirling numbers of the first kind can be interpreted as follows: $(-1)^{n-j}S_n^{(j)}$ *is the number of permutations of a set of n objects having j cycles.*

Recurrence formula (3.34) can also be used to prove an explicit formula, which this time depends on the Stirling numbers of the second kind,

$$S_n^{(j)} = \sum_{k=0}^{n-j}(-1)^k\binom{n-1+k}{n-j+k}\binom{2n-j}{n-j-k}\mathfrak{S}_{n-j+k}^{(k)}.$$

The solution of (3.24) using Stirling numbers then proceeds as follows:

$$\begin{aligned}f_n(k) &= n\sum k^{n-1}\delta k = n\sum\sum_{j=1}^{n-1}\mathfrak{S}_{n-1}^{(j)}[k]_j\delta k \\ &= n\sum_{j=1}^{n-1}\mathfrak{S}_{n-1}^{(j)}\frac{[k]_{j+1}}{j+1} = n\sum_{j=1}^{n-1}\mathfrak{S}_{n-1}^{(j)}\frac{1}{j+1}\sum_{i=1}^{j+1}S_{j+1}^{(i)}k^i \\ &= \sum_{i=1}^{n}k^i\cdot n\sum_{j=i-1}^{n-1}\frac{1}{j+1}\mathfrak{S}_{n-1}^{(j)}S_{j+1}^{(i)} = \sum_{i=1}^{n}k^i\cdot n\sum_{j=i}^{n}\frac{1}{j}\mathfrak{S}_{n-1}^{(j-1)}S_j^{(i)}.\end{aligned}$$

3.5 Renate Golombek's Problem

In 1994 Renate Golombek [Golombek 94] posed the problem of determining the following sums for $r = 1, 2, 3, \ldots$:

$$P(n,r) = \sum_{j=0}^{n}\binom{n}{j}j^r, \quad Q(n,r) = \sum_{j=0}^{n}\binom{n}{j}^2 j^r.$$

The solution uses the Stirling numbers of the second kind (3.31). We apply (3.32) to write

$$P(n,r) = \sum_{k=1}^{r} \mathfrak{S}_r^{(k)} \sum_{j=0}^{n} \binom{n}{j} [j]_k.$$

The inner sum is a binomial sum, cf. (7.1),

$$\sum_{j=0}^{n} \binom{n}{j} [j]_k = [n]_k \sum_{j=k}^{n} \binom{n-k}{j-k} = [n]_k 2^{n-k}.$$

Hence we obtain the formula

$$P(n,r) = \sum_{k=1}^{r} \mathfrak{S}_r^{(k)} [n]_k 2^{n-k}.$$

Proceeding the same way with the squares, we get

$$Q(n,r) = \sum_{k=1}^{r} \mathfrak{S}_r^{(k)} \sum_{j=0}^{n} \binom{n}{j}^2 [j]_k. \tag{3.35}$$

This time we will evaluate the sums

$$\sum_{j=0}^{n} \binom{n}{j}^2 [j]_k.$$

The terms less than $j = k$ vanish, so by reversing the direction of summation we get

$$\sum_{j=0}^{n-k} \binom{n}{j}^2 [n-j]_k.$$

Now we apply formula (2.8) to write this as

$$[n]_k \sum_{j=0}^{n-k} \binom{n}{j} \binom{n-k}{n-k-j}. \tag{3.36}$$

The sum is an example of a *Chu–Vandermonde convolution*, which will be discussed in Chapter 8. By (8.1), it equals

$$[n]_k \binom{2n-k}{n-k}. \tag{3.37}$$

Substitution of (3.37) in (3.35) yields

$$Q(n,r) = \sum_{k=1}^{r} \mathfrak{S}_r^{(k)} [n]_k \binom{2n-k}{n-k}. \tag{3.38}$$

This formula is the general solution to the problem. In order to make it easy to see how much it coincides with the solutions suggested, we will find useful a reformulation of (3.36) using (2.2):

$$[n]_k \binom{2n-k}{n-k} = \binom{2n-r}{n-1} \frac{[2n-k]_{r-k}[n-1]_{k-1}}{[n-k]_{r-k-1}}.$$

With this substituted in (3.38), we get

$$Q(n,r) = \binom{2n-r}{n-1} \sum_{k=1}^{r} \mathfrak{S}_r^{(k)} \frac{[2n-k]_{r-k}[n-1]_{k-1}}{[n-k]_{r-k-1}}. \tag{3.39}$$

Application of (3.39) yields, for $r = 1$,

$$\begin{aligned} Q(n,1) &= \binom{2n-1}{n-1} \frac{[2n-1]_0[n-1]_0}{[n-1]_{-1}} \\ &= \binom{2n-1}{n-1} (n-1+1) \\ &= n\binom{2n-1}{n}. \end{aligned}$$

For $r = 2$, we get

$$\begin{aligned} Q(n,2) &= \binom{2n-2}{n-1} \left(\frac{[2n-1]_1[n-1]_0}{[n-1]_0} + \frac{[2n-1]_0[n-1]_1}{[n-2]_{-1}} \right) \\ &= \binom{2n-2}{n-1} (2n-1+(n-1)^2) \\ &= \binom{2n-2}{n-1} n^2. \end{aligned}$$

And for $r = 3$, we get

$$\begin{aligned} Q(n,3) &= \binom{2n-3}{n-1} \\ &\quad \times \left(\frac{[2n-1]_2[n-1]_0}{[n-1]_1} + 3\frac{[2n-2]_1[n-1]_1}{[n-2]_0} + \frac{[2n-3]_0[n-1]_2}{[n-3]_{-1}} \right) \\ &= \binom{2n-3}{n-1} ((2n-1)2 + 3(2n-2)(n-1) + (n-1)(n-2)^2) \\ &= \binom{2n-3}{n-1} n^2(n+1). \end{aligned}$$

But the pattern does not proceed further. Having just the binomial coefficient in front is too much to ask for in general. Already for $r = 4$ we get a denominator:

$$\begin{aligned} Q(n,4) &= \binom{2n-4}{n-1} \\ &\times \left(\frac{[2n-1]_3}{[n-1]_2} + 7\frac{[2n-2]_2[n-1]_1}{[n-2]_1} + 6[2n-3]_1[n-1]_2 + \frac{[n-1]_3}{[n-4]_{-1}} \right) \\ &= \binom{2n-4}{n-1} n^2 \frac{n^3+n^2-3n-1}{n-2}. \end{aligned}$$

Another way to rewrite (3.38) is to replace the binomial coefficient with factorials:

$$Q(n,r) = \sum_{k=1}^{r \wedge n} \mathfrak{S}_r^{(k)} \frac{[2n-k]_n}{[n-k]_{n-k}}.$$

Chapter 4

Linear Difference Equations

4.1 General Linear Difference Equations

General linear difference equations of order m have the form

$$f(n) = \sum_{k=1}^{m} a_k(n) f(n-k) + g(n), \tag{4.1}$$

where a_k and g are given functions of n, where a_m is not identically zero, and f is wanted. Iteration of (4.1) gives the consecutive values of $f(n)$, $n = m+1, m+2, \ldots$ for given initial values, $f(1), f(2), \ldots, f(m)$. The question is whether it is possible to find a closed form for $f(n)$.

The complete solution is usually divided into a sum of two, the solution to the homogeneous equation, i.e., the equation with $g(n) = 0$, and a particular solution to the equation (4.1) as given.

4.2 The Homogeneous Equation

A set of m solutions $f_1, \ldots, f_m$ to the homogeneous equation (4.1), is called *linearly independent*, if any equation of the form

$$\sum_{j=1}^{m} c_j f_j(n) = 0, \quad n \in \mathbb{Z} \tag{4.2}$$

has coefficients equal to zero, $c_1 = \cdots = c_m = 0$.

In analogy to the Wronskian we can consider the determinant

$$W(n) = \begin{vmatrix} f_1(n-m) & f_2(n-m) & \cdots & f_m(n-m) \\ f_1(n-m+1) & f_2(n-m+1) & \cdots & f_m(n-m+1) \\ \vdots & \vdots & & \vdots \\ f_1(n-1) & f_2(n-1) & \cdots & f_m(n-1) \end{vmatrix}.$$

We can easily compute $W(n)$ by the use of (4.1) with $g = 0$. Row operations yield

$$\begin{aligned} W(n+1) &= \begin{vmatrix} f_1(n-m+1) & f_2(n-m+1) & \cdots & f_m(n-m+1) \\ \vdots & \vdots & & \vdots \\ a_m(n)f_1(n-m) & a_m(n)f_2(n-m) & \cdots & a_m(n)f_m(n-m) \end{vmatrix} \\ &= (-1)^{m-1}a_m(n)W(n). \end{aligned}$$

Hence, if $a_m(n) \neq 0$ for all values of n, then the determinant is either identically zero or never zero. In the latter case, it can be computed as

$$W(n) = (-1)^{(m-1)n}W(0)\prod_{j=0}^{n-1} a_m(j). \tag{4.3}$$

This means that if we start with m independent m-dimensional vectors, $(f_1(1),\ldots,f_1(m)),\ldots,(f_m(1),\ldots,f_m(m))$, the solutions defined by (4.2) will remain independent provided $a_m(n) \neq 0$ for all values of n. Furthermore, any other solution is uniquely defined by its starting values $f(1),\ldots,f(m)$, which are a linear combination of the m vectors above. This shows that the space of solutions to (4.2) is an m-dimensional vector space.

The special case of $m = 1$ is

$$f(n) = a(n)f(n-1). \tag{4.4}$$

We assume that $a(n) \neq 0$, and write the homogeneous equation (4.4) as

$$\frac{f(n)}{f(n-1)} = a(n),$$

giving the solutions for different choices of initial value $f(0)$,

$$f(n) = f(0)\prod_{k=1}^{n} a(k). \tag{4.5}$$

The special case of $m = 2$,

$$f(n) = a_1(n)f(n-1) + a_2(n)f(n-2) \tag{4.6}$$

with two solutions, $f_1(n)$ and $f_2(n)$, gives the determinant

$$W(n) = \begin{vmatrix} f_1(n-2) & f_2(n-2) \\ f_1(n-1) & f_2(n-1) \end{vmatrix} = f_1(n-2)f_2(n-1) - f_2(n-2)f_1(n-1), \tag{4.7}$$

computable by (4.3) as

$$W(n) = (-1)^n W(0) \prod_{j=0}^{n-1} a_2(j).$$

If we know—by guessing perhaps—a solution $f_1(n) \neq 0$, then we can use the determinant to find the other solution $f_2(n)$. If we try the solution $f_2(n) = \phi(n)f_1(n)$ in (4.7), we get

$$f_1(n)\phi(n+1)f_1(n+1) - \phi(n)f_1(n)f_1(n+1) = W(n+2),$$

from which it is easy to find $\phi(n)$ from the form

$$\Delta\phi(n) = \phi(n+1) - \phi(n) = \frac{W(n+2)}{f_1(n)f_1(n+1)}, \tag{4.8}$$

which by definition (1.8) has the solution

$$\phi(n) = \sum \frac{W(n+2)}{f_1(n)f_1(n+1)} \delta n,$$

and hence provides us with the second solution to (4.6):

$$f_2(n) = \phi(n)f_1(n) = f_1(n) \sum \frac{W(n+2)}{f_1(n)f_1(n+1)} \delta n.$$

4.3 First-Order Inhomogeneous Equations

If $m = 1$, we have the form

$$f(n) = a(n)f(n-1) + g(n). \tag{4.9}$$

We assume that $a(n) \neq 0$, and let $f \neq 0$ be any solution of form (4.5) to the homogeneous equation. Then we consider a solution to (4.9) of the form $f\phi$ and get

$$f(n)\phi(n) - a(n)f(n-1)\phi(n-1) = g(n).$$

Together with (4.4) we obtain

$$\phi(n) - \phi(n-1) = \frac{g(n)}{f(n)},$$

which by "telescoping" yields

$$\phi(n) = \sum_{k=1}^{n} \frac{g(k)}{f(k)} + \phi(0),$$

and hence the complete solution to (4.9),

$$f(n)\phi(n) = f(n) \sum_{k=1}^{n} \frac{g(k)}{f(k)} + \phi(0) f(n), \tag{4.10}$$

where the term $\phi(0)f(n)$ is *any* solution to the homogeneous equation.

4.4 First-Order Equations with Constant Coefficients

For $m = 1$ we choose $a(n) = a$ independent of n in (4.9). Then the solution (4.5) to the homogeneous equation becomes

$$f(n) = f(0)a^n, \tag{4.11}$$

giving the complete solution (4.10) of (4.9)

$$\begin{aligned} f(n)\phi(n) &= f(0)a^n \sum_{k=1}^{n} \frac{g(k)}{a^k} + f(0)\phi(0)a^n \\ &= f(0) \sum_{k=1}^{n} g(k)a^{n-k} + f(0)\phi(0)a^n. \end{aligned} \tag{4.12}$$

4.5 Arbitrary-Order Equations with Constant Coefficients

If $m > 1$, consider the following operator, which is a polynomial in the shift operator $\mathbf{E}$ from (1.6):

$$\mathbf{E}^m - \sum_{k=1}^{m} a_k \mathbf{E}^{m-k}. \tag{4.13}$$

Then equation (4.1) takes the form

$$\left(\mathbf{E}^m - \sum_{k=1}^{m} a_k \mathbf{E}^{m-k}\right) f(n) = g(n). \tag{4.14}$$

By the fundamental theorem of algebra, let the characteristic polynomial of (4.13),

$$p(x) = x^m - \sum_{k=1}^{m} a_k x^{m-k}, \tag{4.15}$$

have the distinct complex roots $\alpha_1, \ldots, \alpha_q$ of order $\nu_1, \ldots, \nu_q$, respectively. Then the operator can be split into

$$\mathbf{E}^m - \sum_{k=1}^{m} a_k \mathbf{E}^{m-k} = \prod_{j=1}^{q} (\mathbf{E} - \alpha_j \mathbf{I})^{\nu_j}, \tag{4.16}$$

and equation (4.14) can be solved by repeated use of solution (4.12).

Theorem 4.1. *If the homogeneous equation*

$$\left(\mathbf{E}^m - \sum_{k=1}^{m} a_k \mathbf{E}^{m-k}\right) f(n) = 0 \tag{4.17}$$

can be written in the operator form with distinct α's

$$\prod_{j=1}^{q} (\mathbf{E} - \alpha_j \mathbf{I})^{\nu_j} f(n) = 0, \tag{4.18}$$

then the complete solution is

$$f(n) = \sum_{j=1}^{q} p_j(n) \alpha_j^n, \tag{4.19}$$

where p_j is any polynomial of degree at most $\nu_j - 1$.

Proof: By induction after m. For $m = 1$, (4.19) reduces to (4.11).

Presume the theorem for m, and consider the equation of form

$$\left(\mathbf{E}^m - \sum_{k=1}^{m} a_k \mathbf{E}^{m-k}\right) (\mathbf{E} - \alpha \mathbf{I}) f(n) = 0. \tag{4.20}$$

Then we get the presumed solution

$$(\mathbf{E} - \alpha \mathbf{I}) f(n) = \sum_{j=1}^{q} p_j(n) \alpha_j^n. \tag{4.21}$$

By the linearity of the operator it is enough to solve each of the equations

$$(\mathbf{E}-\alpha\mathbf{I})f(n)=[n]_k\alpha_j^n,\quad k=0,\dots,\nu_j-1;j=0,\dots,q. \tag{4.22}$$

Case 1: $\alpha=\alpha_j$. We put $f(n)=[n]_{k+1}\alpha^{n-1}$. This is allowed because the degree of $\alpha=\alpha_j$ is now ν_j+1. Then we get by (3.3),

$$\begin{aligned}(\mathbf{E}-\alpha\mathbf{I})f(n)&=[n+1]_{k+1}\alpha^n-[n]_{k+1}\alpha^n=\Delta[n]_{k+1}\alpha^n\\&=(k+1)[n]_k\alpha^n,\end{aligned} \tag{4.23}$$

proving that $f(n)=\frac{[n]_{k+1}\alpha^{n-1}}{k+1}$ solves (4.22).

Case 2: $\alpha\neq\alpha_j$. Analogously we try $f(n)=[n]_k\alpha_j^n$. Then we get by (3.3),

$$\begin{aligned}(\mathbf{E}-\alpha\mathbf{I})f(n)&=[n+1]_k\alpha_j^{n+1}-\alpha[n]_k\alpha_j^n\\&=(([n]_k+k[n]_{k-1})\alpha_j-\alpha[n]_k)\alpha_j^n\\&=(\alpha_j-\alpha)[n]_k\alpha_j^n+k[n]_{k-1}\alpha_j^{n+1},\end{aligned} \tag{4.24}$$

showing that $f(n)=\frac{[n]_k\alpha_j^n}{\alpha_j-\alpha}$ solves the problem for $k=0$ and otherwise reduces the problem from k to $k-1$. □

4.6 Systems of Equations with Constant Coefficients

If we define a vector of functions by $f_j(n)=f(n-j+1)$, $j=1,\dots,m$, then the equation (4.14) can be written as a first-order equation in vectors $\mathbf{f}=(f_1,\dots,f_m)$ and $\mathbf{g}(n)=(g(n),0,\dots,0)$,

$$\mathbf{f}(n)=\mathbf{A}\mathbf{f}(n-1)+\mathbf{g}(n), \tag{4.25}$$

with matrix of coefficients

$$\mathbf{A}=\begin{pmatrix}a_1&a_2&a_3&\cdots&a_{m-1}&a_m\\1&0&0&\cdots&0&0\\0&1&0&\cdots&0&0\\\vdots&\vdots&\vdots&&\vdots&\vdots\\0&0&0&\cdots&1&0\end{pmatrix}.$$

In the form (4.25) the 1-dimensional solution can be imitated. The homogeneous system has the solution

$$\mathbf{f}(n)=\mathbf{A}^n\mathbf{f}(0), \tag{4.26}$$

and if a solution to the original system takes the form

$$\mathbf{f}(n) = \mathbf{A}^n \phi(n) \tag{4.27}$$

with any vector function $\phi(n)$, then we get

$$\mathbf{A}^n \phi(n) - \mathbf{A}\mathbf{A}^{n-1}\phi(n-1) = \mathbf{g}(n),$$

which can be written for a regular matrix $\mathbf{A}$ as

$$\phi(n) - \phi(n-1) = \mathbf{A}^{-n}\mathbf{g}(n).$$

By summing it follows that

$$\phi(n) = \sum_{k=1}^{n} \mathbf{A}^{-k}\mathbf{g}(k) + \phi(0),$$

and hence that the solution (4.27) becomes

$$\mathbf{f}(n) = \sum_{k=1}^{n} \mathbf{A}^{n-k}\mathbf{g}(k) + \phi(0)\mathbf{A}^n.$$

The only problem with this solution is that powers of matrices are cumbersome to compute. But we can use the Cayley–Hamilton Theorem 4.2, discussed below, to reduce the computation to the first $m-1$ powers of the matrix $\mathbf{A}$. That is, if the characteristic polynomial of $\mathbf{A}$ is

$$p_{\mathbf{A}}(x) = \det(x\,\mathbf{I} - \mathbf{A}) = x^m + a_{m-1}x^{m-1} + \cdots + a_0,$$

where $\mathbf{I}$ is the unit matrix, then the Cayley–Hamilton Theorem says that $p_{\mathbf{A}}(\mathbf{A}) = \mathbf{O}$, and hence that

$$\mathbf{A}^m = -a_{m-1}\mathbf{A}^{m-1} - \cdots - a_0\,\mathbf{I}.$$

Furthermore, if we apply this equation to the solution (4.26), then we get

$$\mathbf{A}^m\mathbf{f}(n) = -a_{m-1}\mathbf{A}^{m-1}\mathbf{f}(n) - \cdots - a_0\mathbf{f}(n),$$

or better,

$$\mathbf{f}(n+m) = -a_{m-1}\mathbf{f}(n+m-1) - \cdots - a_0\mathbf{f}(n),$$

which equation simply states that each coordinate function of the vector solution to the homogeneous system of equations satisfies the higher order equation with the same characteristic polynomial as the system.

The Cayley–Hamilton Theorem

Let $\mathbf{A}$ be any $n \times n$-matrix, $\mathbf{A} = (a_{ij})$. Then the determinant can be computed by the development after a column or row, e.g.,

$$\det \mathbf{A} = \sum_{j=1}^{n} a_{ij}(-1)^{i+j} \det \mathbf{A}^{(i,j)},$$

where $\mathbf{A}^{(i,j)}$ is the (i,j)-th complement, i.e., the matrix obtained by omitting the i-th row and the j-th column from the matrix $\mathbf{A}$. This means that if we define a matrix $\mathbf{B} = (b_{jk})$ as

$$b_{jk} = (-1)^{k+j} \det \mathbf{A}^{(k,j)},$$

then we have traced the inverse of the matrix $\mathbf{A}$, provided it exists. At least we can write

$$\sum_{j=1}^{n} a_{ij} b_{jk} = \delta_{ik} \det \mathbf{A},$$

or in matrix form,

$$\mathbf{AB} = (\det \mathbf{A})\,\mathbf{I}. \tag{4.28}$$

Theorem 4.2 (Cayley–Hamilton Theorem). *If*

$$p(\xi) = \det(\xi\,\mathbf{I} - \mathbf{A}) = \xi^n + a_{n-1}\xi^{n-1} + \cdots + a_0 \tag{4.29}$$

is the characteristic polynomial for the matrix $\mathbf{A}$, *then the matrix* $p(\mathbf{A})$ *satisfies*

$$p(\mathbf{A}) = \mathbf{A}^n + a_{n-1}\mathbf{A}^{n-1} + \cdots + a_0\,\mathbf{I} = \mathbf{O}. \tag{4.30}$$

Proof: We apply (4.28) to the matrix $\lambda\,\mathbf{I} - \mathbf{A}$ to get

$$p(\lambda)\,\mathbf{I} = (\lambda\,\mathbf{I} - \mathbf{A})\,\mathbf{B}(\lambda), \tag{4.31}$$

where $\mathbf{B}(\lambda) = (b_{ij}(\lambda))$ is a matrix of polynomials in λ, defined as the (j,i)-th complement of the matrix $\lambda\,\mathbf{I} - \mathbf{A}$. Hence we can write

$$\mathbf{B}(\lambda) = \lambda^{n-1}\mathbf{B}_{n-1} + \cdots + \lambda\mathbf{B}_1 + \mathbf{B}_0$$

as a polynomial in λ with coefficients which are matrices independent of λ.

For any $k \geq 1$ we can write

$$\mathbf{A}^k - \lambda^k\,\mathbf{I} = (\mathbf{A} - \lambda\,\mathbf{I})\left(\mathbf{A}^{k-1} + \lambda\mathbf{A}^{k-2} + \cdots + \lambda^{k-1}\mathbf{I}\right).$$

Hence we can write

$$\begin{aligned} &p(\mathbf{A}) - p(\lambda)\,\mathbf{I} \\ &\quad = \mathbf{A}^n - \lambda^n\,\mathbf{I} + a_{n-1}\left(\mathbf{A}^{n-1} - \lambda^{n-1}\,\mathbf{I}\right) + \cdots + a_1\left(\mathbf{A}^1 - \lambda^1\,\mathbf{I}\right) \\ &\quad = (\mathbf{A} - \lambda\,\mathbf{I})\,\mathbf{C}(\lambda) \\ &\quad = (\mathbf{A} - \lambda\,\mathbf{I})\left(\lambda^{n-1}\,\mathbf{I} + \lambda^{n-2}\mathbf{C}_{n-2} + \cdots + \mathbf{C}_0\right), \end{aligned} \tag{4.32}$$

where $\mathbf{C}(\lambda)$ is a polynomial in λ with coefficients which are matrices independent of λ.

Adding (4.31) and (4.32) we get

$$\begin{aligned} p(\mathbf{A}) &= (\mathbf{A} - \lambda\,\mathbf{I})\,(\mathbf{C}(\lambda) - \mathbf{B}(\lambda)) \\ &= (\mathbf{A} - \lambda\,\mathbf{I})\left(\lambda^{n-1}\,(\mathbf{I} - \mathbf{B}_{n-1}) + \cdots + \lambda\,(\mathbf{C}_1 - \mathbf{B}_1) + \mathbf{C}_0 - \mathbf{B}_0\right) \\ &= \lambda^{k+1}\,(\mathbf{B}_k - \mathbf{C}_k) + \lambda^k \cdots, \end{aligned}$$

where k is the degree of the second factor to the right. This polynomial in λ can only be constant, i.e., independent of λ, if the second factor is zero. But in that case it is all zero, or $p(\mathbf{A}) = \mathbf{O}$. □

4.7 Generating Functions

The equation (4.1) with $g = 0$,

$$f(n+m) = \sum_{k=0}^{m-1} a_k f(n+k) \tag{4.33}$$

can be solved by defining the series

$$F(x) = \sum_{n=0}^{\infty} \frac{x^n}{n!} f(n),$$

called the *exponential generating function* for (4.33). It satisfies

$$F^{(k)}(x) = \sum_{n=k}^{\infty} \frac{x^{n-k}}{(n-k)!} f(n) = \sum_{n=0}^{\infty} \frac{x^n}{n!} f(n+k).$$

Hence we get

$$\sum_{k=0}^{m-1} a_k F^{(k)}(x) = \sum_{n=0}^{\infty} \frac{x^n}{n!} \sum_{k=0}^{m-1} a_k f(n+k) = \sum_{n=0}^{\infty} \frac{x^n}{n!} f(n+m) = F^{(m)}(x).$$

This is a differential equation we can solve.

For instance, take the difference equation

$$f(2+n) = f(1+n) + f(n).$$

The exponential generating function satisfies the equation

$$F'' = F' + F,$$

with the roots $\xi_\pm = \frac{1 \pm \sqrt{5}}{2}$, so that F becomes a combination of

$$e^{\xi_\pm x} = \sum_{n=0}^{\infty} \xi_\pm^n \frac{x^n}{n!}.$$

Even the Bernoulli polynomials have an exponential generating function:

$$\frac{xe^{kx}}{e^x - 1} = \sum_{n=0}^{\infty} \frac{B_n(k)x^n}{n!}. \tag{4.34}$$

Consider the product

$$e^x \sum_{n=0}^{\infty} \frac{B_n(k)x^n}{n!} = \sum_{n=0}^{\infty} \frac{x^n}{n!} \sum_{j=0}^{n} \binom{n}{j} B_j(k).$$

Now from (3.30) this is equal to

$$\sum_{n=0}^{\infty} \frac{x^n}{n!} \left(nk^{n-1} + B_n(k)\right) = xe^{kx} + \sum_{n=0}^{\infty} \frac{B_n(k)x^n}{n!}.$$

4.8 Chen Kwang-Wu's Problem

In 1994, Chen Kwang-Wu [Chen 94] posed the problem of proving the following identity for Bernoulli polynomials with $m \geq 1$:

$$\sum_{k=0}^{m} \binom{m}{k} B_k(\alpha) B_{m-k}(\beta) = -(m-1)B_m(\alpha+\beta) + m(\alpha+\beta-1)B_{m-1}(\alpha+\beta). \tag{4.35}$$

Proof: Multiplication of two copies of (4.34), taken for $t = x$, $k = \alpha$, and $k = \beta$ respectively, yields

$$\frac{t^2 e^{(\alpha+\beta)t}}{(e^t-1)^2} = \sum_{n=0}^{\infty} \frac{t^n}{n!} \sum_{k=0}^{n} \binom{n}{k} B_k(\alpha) B_{n-k}(\beta). \tag{4.36}$$

By differentiation of (4.34) with respect to t and then multiplication with t, we also have the formula

$$\frac{te^{xt}}{e^t-1} + xt\frac{te^{xt}}{e^t-1} - \frac{t^2 e^{(x+1)t}}{(e^t-1)^2} = \sum_{n=0}^{\infty} \frac{nB_n(x)t^n}{n!}.$$

Substitution of the series from (4.34) in this formula with the choice of $x = \alpha + \beta - 1$ yields

$$\begin{aligned}\frac{t^2 e^{(\alpha+\beta)t}}{(e^t-1)^2} = 1 + \sum_{n=1}^{\infty} \frac{t^n}{n!} \Big(&- (n-1) B_n(\alpha+\beta-1) \\ &+ n(\alpha+\beta-1) B_{n-1}(\alpha+\beta-1) \Big).\end{aligned} \tag{4.37}$$

From the same difference equation which the Bernoulli polynomials were invented to solve, i.e., (3.24), we get

$$B_m(\alpha+\beta-1) = B_m(\alpha+\beta) - m(\alpha+\beta-1)^{m-1}$$

for $m = n$ and $m = n-1$, respectively in (4.37), and the formula (4.35) follows from comparing the terms in (4.37) and (4.36). □

4.9 Equations with Polynomial Coefficients

If in (4.1) the functions $a_k(n)$ are polynomials in n, certain higher order equations can be solved. It is always possible to transform the equation to a differential equation, not necessarily of the same order as the difference equation.

The simplest case is the difference equation of order 2 with coefficients of order 1:

$$(c_0+b_0 n)f(n) + (c_1+b_1(n-1))f(n-1) + (c_2+b_2(n-2))f(n-2) = 0. \tag{4.38}$$

One approach consists in the introduction of the *generating function for the solution* $f(n)$,

$$F(x) = \sum_{n=0}^{\infty} f(n) x^n.$$

This function obeys two simple rules of our concern, written conveniently at once as

$$\begin{aligned} x^k F^{(j)}(x) &= \sum_{n=j}^{\infty} x^k f(n)[n]_j x^{n-j} \\ &= \sum_{n=j}^{\infty} [n]_j f(n) x^{n+k-j} \\ &= \sum_{n=k}^{\infty} [n-k+j]_j f(n-k+j) x^n. \end{aligned}$$

To handle (4.38) it is enough to consider $j = 0, 1$, hence the differential equation gets order 1, and to consider $k = 0, 1, 2, 3$ to obtain the shift of order 2. The differential equation in F becomes

$$\left(c_0 + c_1 x + c_2 x^2\right) F(x) + \left(b_0 x + b_1 x^2 + b_2 x^3\right) F'(x) = 0. \tag{4.39}$$

If we find the zeros of the polynomial $b_0 + b_1 x + b_2 x^2$, say a_1, a_2, then we can use partial fractions to write the differential equation (4.39) in general as

$$\frac{F'(x)}{F(x)} = -\frac{c_0 + c_1 x + c_2 x^2}{b_0 x + b_1 x^2 + b_2 x^3} = \frac{\beta_0}{x} + \frac{\beta_1}{x - a_1} + \frac{\beta_2}{x - a_2}.$$

In the cases of double roots and common zeros, the expressions are similar. We note that the constant β_0 is equal to $\frac{b_2 c_0}{b_0}$. This can be solved as

$$F(x) = x^{\beta_0} (x - a_1)^{\beta_1} (x - a_2)^{\beta_2}.$$

If β_0 is an integer, we can develop the other factors as series in x, cf. (7.2),

$$(x - a)^{\beta} = \sum_{k=0}^{\infty} \binom{\beta}{k} (-a)^{\beta - k} x^k,$$

and take their product to find

$$\begin{aligned} F(x) &= x^{\beta_0} \sum_{n=0}^{\infty} x^n \sum_{k=0}^{n} \binom{\beta_1}{k} \binom{\beta_2}{n-k} (-a_1)^{\beta_1 - k} (-a_2)^{\beta_2 + k - n} \\ &= \sum_{n=0}^{\infty} x^n \sum_{k=0}^{n-\beta_0} \binom{\beta_1}{k} \binom{\beta_2}{n - \beta_0 - k} (-a_1)^{\beta_1 - k} (-a_2)^{\beta_2 + k - n + \beta_0}, \end{aligned}$$

yielding the solution to equation (4.38) as

$$f(n) = \sum_{k=0}^{n-\beta_0} \binom{\beta_1}{k} \binom{\beta_2}{n - \beta_0 - k} (-a_1)^{\beta_1 - k} (-a_2)^{\beta_2 + k - n + \beta_0}. \tag{4.40}$$

In order to write this solution in the forms treated in this book, we prefer to rewrite (4.40) as

$$f(n) = (-a_1)^{\beta_1}(-a_2)^{\beta_2 - n + \beta_0} \sum_{k=0}^{n-\beta_0} \binom{n-\beta_0}{k} [\beta_1]_k [\beta_2]_{n-\beta_0-k} \left(\frac{a_2}{a_1}\right)^k .$$

4.10 Shalosh B. Ekhad's Squares

In 1994 Shalosh B. Ekhad [Ekhad 94] posed the following problem:

> *Let X_n be defined by $X_0 = 0, X_1 = 1, X_2 = 0, X_3 = 1$, and for $n \geq 1$,*
>
> $$X_{n+3} = \frac{(n^2+n+1)(n+1)}{n} X_{n+2} + (n^2+n+1) X_{n+1} - \frac{n+1}{n} X_n.$$
>
> *Prove that X_n is the square of an integer for every $n \geq 0$.*

We consider the general second-order difference equation,

$$x_{n+2} = f(n)x_{n+1} + g(n)x_n \qquad f, g : \mathbb{N} \to \mathbb{Z} \setminus \{0\}. \tag{4.41}$$

With any integral initial values, x_1, x_2, this equation will generate a sequence of integers. We will see that, with the suitable choice of functions and initial values, the solution to (4.41) will be the sequence of square roots asked for in the problem.

We consider the equation (4.41) in the two forms

$$\begin{aligned} x_{n+3} &= f(n+1)x_{n+2} + g(n+1)x_{n+1}, \\ g(n)x_n &= x_{n+2} - f(n)x_{n+1}. \end{aligned}$$

By squaring these two equations and then eliminating the mixed products $x_{n+2}x_{n+1}$, we get a third-order difference equation in the squares:

$$\frac{1}{f(n+1)g(n+1)} x_{n+3}^2 = \left(\frac{f(n+1)}{g(n+1)} + \frac{1}{f(n)}\right) x_{n+2}^2 + \left(\frac{g(n+1)}{f(n+1)} + f(n)\right) x_{n+1}^2 - \frac{g(n)^2}{f(n)} x_n^2.$$

With the choice of the functions $f(n) = n,\ g(n) = 1$, we obtain an equation for $X_n = x_n^2$, equivalent to the one posed in the problem:

$$\frac{1}{n+1} X_{n+3} = \left((n+1) + \frac{1}{n}\right) X_{n+2} + \left(\frac{1}{n+1} + n\right) X_{n+1} - \frac{1}{n} X_n.$$

With the initial values $x_1 = 1$, $x_2 = 0$ we obtain $x_3 = 1$ and the initial values of the problem (the value of X_0 is irrelevant, provided it is a square). Hence the solution to the posed problem consists of integral squares.

4.11 David Doster's Problem

In 1994, David Doster [Doster 94] posed the following problem:

Define a sequence $\langle y_n \rangle$ *recursively by* $y_0 = 1$, $y_1 = 3$, *and*

$$y_{n+1} = (2n+3)y_n - 2ny_{n-1} + 8n$$

for $n \geq 1$. *Find an asymptotic formula for* y_n.

Solution.

$$y_n \sim 2^{n+1} n! \sqrt{e},$$

while an exact expression for y_n is

$$y_n = 1 + 2n + \sum_{k=2}^{n} [n]_k 2^{k+1}.$$

Proof: We define a new sequence, z_n, by the formula

$$y_n = 2^n n! z_n.$$

Then this sequence is defined by $z_0 = 1$, $z_1 = \frac{3}{2}$, and the recursion

$$z_{n+1} = z_n + \frac{z_n - z_{n-1}}{2(n+1)} + \frac{4n}{2^n(n+1)!}$$

for $n \geq 1$. This is really a difference equation in the difference

$$x_n = \Delta z_n = z_{n+1} - z_n,$$

which sequence is defined by $x_0 = \frac{1}{2}$ and for $n \geq 1$ the recursion

$$x_n = \frac{x_{n-1}}{2(n+1)} + \frac{4n}{2^n(n+1)!},$$

with the straightforward solution for $n \geq 1$,

$$x_n = \frac{1}{2^{n-1}(n-1)!} + \frac{1}{2^{n+1}(n+1)!}.$$

Hence we find

$$\begin{aligned} z_n &= z_0 + \sum_{k=0}^{n-1} x_k = 1 + \tfrac{1}{2} + \sum_{k=1}^{n-1} \frac{1}{2^{k-1}(k-1)!} + \sum_{k=1}^{n-1} \frac{1}{2^{k+1}(k+1)!} \\ &= \sum_{k=0}^{n-2} \frac{\left(\frac{1}{2}\right)^k}{k!} + 1 + \tfrac{1}{2} + \sum_{k=2}^{n} \frac{\left(\frac{1}{2}\right)^k}{k!} = \sum_{k=0}^{n-2} \frac{\left(\frac{1}{2}\right)^k}{k!} + \sum_{k=0}^{n} \frac{\left(\frac{1}{2}\right)^k}{k!}. \end{aligned}$$

It follows that the limit must be

$$\lim_{n\to\infty} z_n = 2e^{\frac{1}{2}},$$

and that we can compute y_n as

$$\begin{aligned} y_n &= 2^n n! \left(\sum_{k=0}^{n-2} \frac{\left(\frac{1}{2}\right)^k}{k!} + \sum_{k=0}^{n} \frac{\left(\frac{1}{2}\right)^k}{k!} \right) \\ &= \sum_{k=0}^{n-2} 2^{n-k} [n]_{n-k} + \sum_{k=0}^{n} 2^{n-k} [n]_{n-k} \\ &= 1 + 2n + \sum_{k=2}^{n} 2^{k+1} [n]_k. \end{aligned}$$

□

4.12 Ira Gessel's Problem

In 1995 Ira Gessel [Gessel 95b, Andersen and Larsen 97] posed this problem:

Evaluate the following sum for all $n \in \mathbb{N}$:

$$S(n) = \sum_{3k\le n} 2^k \frac{n}{n-k} \binom{n-k}{2k}. \tag{4.42}$$

Solution. For all $n \in \mathbb{N}$, we have

$$S(n) = \sum_{3k\le n} 2^k \frac{n}{n-k} \binom{n-k}{2k} = 2^{n-1} + \cos\left(n \cdot \tfrac{\pi}{2}\right). \tag{4.43}$$

Proof: The function $S(n)$ defined by (4.42) satisfies the difference equation

$$S(n) - 2S(n-1) + S(n-2) - 2S(n-3) = 0. \tag{4.44}$$

Hence the solution must take the form

$$S(n) = \alpha 2^n + \beta i^n + \gamma(-i)^n.$$

Setting $S(1) = 1$, $S(2) = 1$, and $S(3) = 4$ yields $\alpha = \beta = \gamma = \frac{1}{2}$.

In order to establish (4.44) it is convenient to split the sum $S(n)$ into two parts, $S(n) = U(n) + V(n)$, where

$$U(n) = \sum_{3k \le n} 2^k \binom{n-k}{2k}, \quad V(n) = \sum_{3k \le n-3} 2^k \binom{n-2-k}{2k+1}.$$

After splitting the binomial coefficients in the sums, these two functions obviously satisfy the simultaneous difference equations

$$\begin{aligned} U(n+1) - U(n) &= 2V(n+1), \\ V(n+1) - V(n) &= U(n-2). \end{aligned}$$

From these it follows that both $U(n)$ and $V(n)$ satisfy the difference equation (4.44), and hence so does their sum $S(n)$. □

Remark 4.3. The solution for $U(n)$ is somewhat ugly:

$$U(n) = \tfrac{1}{5}\left(2^{n+1} + 3\cos\left(n \cdot \tfrac{\pi}{2}\right) + \sin\left(n \cdot \tfrac{\pi}{2}\right)\right),$$

while for $V(n)$ it is similar:

$$V(n) = \tfrac{1}{5}\left(2^{n-1} + 2\cos\left(n \cdot \tfrac{\pi}{2}\right) - \sin\left(n \cdot \tfrac{\pi}{2}\right)\right).$$

It is seen that adding the two gives the simpler form (4.43).

4.13 Emre Alkan's Problem

In 1995 Emre Alkan [Alkan 95] posed the following problem:

Prove there are infinitely many positive integers m such that

$$\frac{1}{5 \cdot 2^m} \sum_{k=0}^{m} \binom{2m+1}{2k} 3^k$$

is an odd integer.

Solution. This expression is not an integer for all non-negative arguments, $m \in \mathbb{N}_0$, so we will consider the function of m,

$$f(m) = \frac{1}{2^m} \sum_{k=0}^{m} \binom{2m+1}{2k} 3^k, \tag{4.45}$$

which function will be proved to be a sequence of odd integers, sometimes divisible by 5.

The sum in (4.45) is half of a binomial sum, so let us add and subtract the remaining terms and write

$$f(m) = \frac{1}{2^{m+1}} \sum_{k=0}^{2m+1} \binom{2m+1}{k} \left(\left(\sqrt{3}\right)^k + \left(-\sqrt{3}\right)^k \right).$$

Then we can apply the binomial theorem (7.1) twice to this and get

$$f(m) = \frac{1}{2^{m+1}} \left(\left(1+\sqrt{3}\right)^{2m+1} + \left(1-\sqrt{3}\right)^{2m+1} \right).$$

Now distribute the 2's to get a sum of two plain powers:

$$f(m) = \frac{1+\sqrt{3}}{2} \left(\frac{\left(1+\sqrt{3}\right)^2}{2} \right)^m + \frac{1-\sqrt{3}}{2} \left(\frac{\left(1-\sqrt{3}\right)^2}{2} \right)^m.$$

Hence the function f must satisfy a difference equation with the two roots $\frac{(1\pm\sqrt{3})^2}{2}$, i.e., with the characteristic polynomial

$$x^2 - 4x + 1,$$

giving the difference equation

$$f(m+2) = 4f(m+1) - f(m).$$

As we easily have $f(0) = 1, f(1) = 5$, which are odd integers, all of the values of f must be odd integers. Hence, if there are some divisible by 5, the quotient must be odd for each of them.

Rather than finding other values divisible by 5, we will prove the following property, from which it follows that there are infinitely many such values.

Theorem 4.4. *For any integer $z \in \mathbb{Z}$ the sums of integers for $m \in \mathbb{N}_0$,*

$$S_m(z) = \sum_{k=0}^{m} \binom{2m+1}{2k} z^k \tag{4.46}$$

satisfy that for all $m \in \mathbb{N}_0$, $S_m(z)$ divides $S_{3m+1}(z)$.

Proof: For any real or complex square root of z we have

$$S_m(z) = \frac{1}{2}\sum_{k=0}^{2m+1}\binom{2m+1}{k}\left(\left(\sqrt{z}\right)^k + \left(-\sqrt{z}\right)^k\right)$$
$$= \frac{1}{2}\left(\left(1+\sqrt{z}\right)^{2m+1} + \left(1-\sqrt{z}\right)^{2m+1}\right). \tag{4.47}$$

Using the identity

$$x^3 + y^3 = (x+y)\left(x^2 - xy + y^2\right), \tag{4.48}$$

we get from (4.47) that

$$S_{3m+1}(z) = \frac{1}{2}\left(\left(1+\sqrt{z}\right)^{6m+3} + \left(1-\sqrt{z}\right)^{6m+3}\right)$$
$$= \frac{1}{2}\left(\left(1+\sqrt{z}\right)^{2m+1} + \left(1-\sqrt{z}\right)^{2m+1}\right)$$
$$\times\left(\left(1+\sqrt{z}\right)^{4m+2} - \left(1+\sqrt{z}\right)^{2m+1}\left(1-\sqrt{z}\right)^{2m+1} + \left(1-\sqrt{z}\right)^{4m+2}\right)$$
$$= S_m(z)\left(\left(1+\sqrt{z}\right)^{4m+2} - (1-z)^{2m+1} + \left(1-\sqrt{z}\right)^{4m+2}\right),$$

where the second factor in (4.48) is an integer, since the odd powers of the square roots must cancel. □

From this theorem we get that $5 = f(1)$ is a divisor of $f(4)$, which divides $f(13)$, etc.

Observe that for any $j \in \mathbb{N}$ the sequence (4.46) satisfies the difference equation

$$S_{m+2j}(z) = 2\sum_i \binom{j}{2i}(z+1)^{j-2i}(4z)^i\, S_{m+j}(z) - (z-1)^{2j}\, S_m(z). \tag{4.49}$$

So with $j = 2m+1$, we see that each term of the sequence $S_m(z)$, $S_{3m+1}(z)$, $S_{5m+2}(z)$ etc., is divisible by the first term $S_m(z)$. Then with $z = 3$, it follows that 5 is a divisor of $f(7)$, $f(10)$, etc.

The divisor 5 can be replaced by any divisor in any term of the sequence, but what about 3 in 3^k? Of course, it can be replaced by any integer, $z \in \mathbb{Z}$, and for odd integers the sum (4.46) becomes divisible by a power of 2. We actually get three cases: z even, $z \equiv 3$ (4), and $z \equiv 1$ (4).

The first result is that for z even we have that all values are odd integers and the sequence satisfies the difference equation

$$S_{m+2}(z) = 2(z+1)S_{m+1}(z) - (z-1)^2 S_m(z),$$

easily derived from (4.47), and the start values $S_0(z) = 1$ and $S_2(z) = 1 + 3z$, easily derived from (4.46).

The second result is that for $3 + 4z$, the sum

$$U_m(z) = \frac{1}{2^m} \sum_{k=0}^{m} \binom{2m+1}{2k} (3+4z)^k$$

satisfies the difference equation

$$U_{m+2}(z) = 4(z+1)U_{m+1}(z) - (2z+1)^2 U_m(z),$$

and the start values are $U_0(z) = 1$ and $U_1(z) = 5 + 6z$. So in this case we get only odd integers.

And the third result is that for $1 + 4z$, the sum

$$T_m(z) = \frac{1}{4^m} \sum_{k=0}^{m} \binom{2m+1}{2k} (1+4z)^k$$

satisfies the difference equation

$$T_{m+2}(z) = (1+2z)T_{m+1}(z) - z^2 T_m(z),$$

and the start values are $T_0(z) = 1$ and $T_1(z) = 1 + 3z$. If z is even, then the terms are all odd, but if z is odd, they must alternate with a period of length 3.

In the binomial coefficients $\binom{2m+1}{2k}$, we can replace $2m + 1$ with $2m$ or $2k$ by $2k + 1$, e.g., $\binom{2m}{2k+1}$, and obtain similar results.

Chapter 5

Classification of Sums

This classification is due to the late Erik Sparre Andersen (1919–2003) [Andersen 89]. We consider sums of the form

$$T(c,n) = \sum_{k=0}^{n} t(c,n,k), \tag{5.1}$$

with $n \in \mathbb{N}_0$ the limit of summation, $c \in \mathbb{C}^\ell$ an argument-vector, and $k \in \mathbb{N}_0$ the summation variable. We can assume the lower limit to be equal to zero, because otherwise it is trivial to translate the formula to obtain this limit.

The question is, do we know a formula which by a trivial transformation gives us the value of the sum (5.1)? By "trivial" we mean obtained by the following three operations:

(1) multiplying with a non-zero constant which can depend on c and n,
(2) special choice of the arguments and limits,
(3) reversing the direction of summation.

The way to treat the first triviality is to consider the *formal quotient*

$$q_t(c,n,k) := \frac{t(c,n,k+1)}{t(c,n,k)}, \tag{5.2}$$

which characterizes the expressions up to proportionality, because of the formula

$$T(c,n) = t(c,n,0) \sum_{k=0}^{n} \prod_{i=0}^{k-1} q_t(c,n,i). \tag{5.3}$$

Hence, it is tempting to classify mainly according to the character of the quotient (5.2).

5.1 Classification

We will classify the sums (5.1) according to the nature of the terms and the quotient (5.3) in five types numbered I–V. If the quotients are independent of the limit n, we call the sum *indefinite*, otherwise *definite*.

We shall apply the main classification as follows:

I. The terms take the form

$$t(c,n,k) = r(c,n,k)z^k, \tag{5.4}$$

where $r(c,n,k)$ is a rational function of k. Typical examples are the quotient series (3.1),

$$\sum_{k=0}^{n} z^k = \begin{cases} \dfrac{z^{n+1}-1}{z-1} & (z \neq 1), \\ n+1 & (z=1), \end{cases}$$

and the sum of polynomials, e.g., (2.6),

$$\sum\nolimits_0^n [k]_m \delta k = \begin{cases} \dfrac{[n]_{m+1}}{m+1} & \text{for } m \neq -1, \\ H_n & \text{for } m = -1. \end{cases}$$

II. Not of type I, but the quotients (5.2) are rational functions of k. A typical example is the binomial formula (7.1),

$$\sum_{k=0}^{n} \binom{n}{k} z^k = (1+z)^n,$$

with quotient equal to

$$q(z,n,k) = \frac{n-k}{-1-k}(-z).$$

III. Not of type I or II, but the terms are products or quotients of terms which are of type I or II by factors or divisors of the form $[x+ky]_k$, where $0 \neq y \neq \pm 1$. A typical example is the Hagen–Rothe formula (13.5),

$$\sum_{k=0}^{n} \binom{n}{k} [x+kz-1]_{k-1}[y+(n-k)z-1]_{n-k-1}$$
$$= \frac{x+y}{xy}[x+y+nz-1]_{n-1}.$$

IV. Not of type I or II, but the terms are products or quotients of terms which are of type I or II, by factors or divisors of the form $(x+ky)^k$, where $y \neq 0$. A typical example is the Abel formula (13.6),

$$\sum_{k=0}^{n} \binom{n}{k} (x+kz)^{k-1}(y-kz)^{n-k-1} = \frac{x+y-nz}{x(y-nz)}(x+y)^{n-1}.$$

V. Not of type I or II, but the terms are products of a term of type I or II and a harmonic number H_k or $H_{c,k}^{(m)}$, cf. (1.11)–(1.12). A typical example is the formula

$$\sum_{k=0}^{n} (-1)^{k-1} \binom{n}{k} H_k = \frac{1}{n}.$$

5.2 Canonical Forms of Sums of Types I–II

In order to recognize a sum, it is convenient to write it in a standard form. By definition the quotients (5.2) are rational functions of k, so they must take the form

$$q_t(c,n,k) = \frac{(\alpha_1 - k)\cdots(\alpha_p - k)}{(\beta_1 - k)\cdots(\beta_q - k)} z, \tag{5.5}$$

with $z \neq 0$ and $\alpha_i \neq \beta_j$ independent of k, but depending on c and n.

The type of the sum is said to have the *parameters* (p,q,z). So we say a sum is of type II(p,q,z) if the quotient is as in (5.5), provided it is not of type I.

The products of quotients in (5.3) can by use of (5.5) be written as

$$r(k) := \frac{\prod_{j=1}^{p} [\alpha_j]_k}{\prod_{j=1}^{q} [\beta_j]_k} z^k. \tag{5.6}$$

The function $r(k)$ in (5.6) can vanish for certain values of k if $\alpha_j \in \mathbb{N}_0$ or $-\beta_j \in \mathbb{N}$. Similarly, it can be undefined or infinity if $-\alpha_j \in \mathbb{N}$ or $\beta_j \in \mathbb{N}_0$.

Definition 5.1. We call a sum (5.3) with quotient of the form (5.5) a *sum of natural limits* if the function $t(c,n,k)$ is defined for all $0 \leq k \leq n$ and $\beta_j = -1$, $\alpha_i = n$ for some (i,j).

In this case we write

$$\frac{[\alpha_i]_k}{[\beta_j]_k} = \frac{[n]_k}{[-1]_k} = \binom{n}{k}(-1)^k.$$

Without loss of generality, we can assume that $i = j = 1$. If we also have the situation that $\beta_j < 0$ or $n \le \beta_j$ for $j = 1, 2, \ldots, q$, then we can apply

$$\frac{1}{[\beta_j]_k} = \frac{[\beta_j - k]_{n-k}}{[\beta_j]_n} = \frac{[n-1-\beta_j]_{n-k}(-1)^{n-k}}{[\beta_j]_n} = \frac{[b_j]_{n-k}(-1)^{n-k}}{[\beta_j]_n}, \quad (5.7)$$

to replace the denominator $[\beta_j]_k$ by the numerator $[b_j]_{n-k}(-1)^{n-k}$, where we have set $b_j = n - 1 - \beta_j$.

With the replacements $a_j = \alpha_j$ and $x = (-1)^q z$ we can define the following.

Definition 5.2. The *canonical form* of a sum of natural limits (5.3) with quotient of the form (5.5) is

$$S(c,n) = \sum_{k=0}^{n} s(c,n,k) = \sum_{k=0}^{n} \binom{n}{k} \prod_{j=2}^{p} [a_j]_k \prod_{j=2}^{q} [b_j]_{n-k}\, x^k. \quad (5.8)$$

From (5.8) it is easy to find the desired sum from (5.3) as

$$T(c,n) = \frac{t(c,n,0)}{s(c,n,0)} S(c,n). \quad (5.9)$$

If the difference between a pair of roots equals one, say $\beta_2 - \alpha_2 = 1$, i.e., $a_2 + b_2 = n - 2$, we can replace the corresponding product by

$$\begin{aligned}[a_2]_k\,[b_2]_{n-k} &= [a_2]_{k-1}\,(a_2 - k + 1)\,[n - a_2 - 2]_{n-k} \\ &= (a_2 + 1 - k)\,[a_2]_{k-1}\,[a_2 - k + 1]_{n-k}\,(-1)^{n-k} \\ &= [a_2]_{n-1}\,(-1)^n\,(a_2 + 1 - k)\,(-1)^k.\end{aligned}$$

The constant terms can be ignored, so we define the following.

Definition 5.3. The *special canonical form* of a sum having a pair of roots with difference 1, i.e., $a_2 + b_2 = n - 2$ or $\beta_2 - \alpha_2 = 1$, is

$$S(c,n) = \sum_{k=0}^{n} s(c,n,k) = \sum_{k=0}^{n} \binom{n}{k} \prod_{j=3}^{p} [a_j]_k \prod_{j=3}^{q} [b_j]_{n-k}\,(a_2 + 1 - k)\ (-x)^k.$$

Remark 5.4. In the form (5.8), it is particularly easy to change the direction of summation, as the sum must be equal to

$$S(c,n) = \sum_{k=0}^{n} \binom{n}{k} \prod_{j=2}^{q} [b_j]_k \prod_{j=2}^{p} [a_j]_{n-k} \left(\frac{1}{x}\right)^k,$$

while the roots are interchanged between numerator and denominator, with the b_j's in the numerator and the $n - 1 - a_j$'s in the denominator.

5.3 Sums of Arbitrary Limits

We can try to reconstruct the sums of form (5.1) from the quotients (5.5) for all possible limits $n \in \mathbb{N}$. The problem is that we have difficulties summing past the integral roots of the rational function (5.5).

For the cases when there are no roots in the interval $[0, n)$, the obvious idea is to sum the products. To deal with other cases, suppose we have two integral roots, say α and β, satisfying

$$0 \leq \alpha < \beta < n,$$

and that we have one of the two cases

$$t((\alpha, \beta), n, k) = \begin{cases} [\beta - k]_{\beta-\alpha}, \\ [k - \alpha - 1]_{\beta-\alpha}. \end{cases}$$

Then in both cases the term factor becomes

$$\frac{t((\alpha, \beta), n, k)}{t((\alpha, \beta), n, 0)} = g(k) = \frac{[\beta - k]_{\beta-\alpha}}{[\beta]_{\beta-\alpha}},$$

which gives the quotient

$$q_g(k) = \frac{[\beta - k - 1]_{\beta-\alpha}}{[\beta - k]_{\beta-\alpha}} = \frac{\alpha - k}{\beta - k}.$$

But, we must admit that

$$g(k) = 0 \quad \text{for } \alpha < k \leq \beta.$$

The natural question is, are there other non-vanishing term factors giving the same quotient inside the interval $(\alpha, \beta]$? The answer is yes; we may consider

$$h(k) = \frac{(-1)^k}{\binom{\beta-\alpha-1}{k-\alpha-1}},$$

which is defined for $\alpha < k \leq \beta$ and gives the desired quotient,

$$\begin{aligned} q_h(k) &= -\frac{\binom{\beta-\alpha-1}{k-\alpha-1}}{\binom{\beta-\alpha-1}{k-\alpha}} = -\frac{[\beta - \alpha - 1]_{k-\alpha-1}[k - \alpha]_{k-\alpha}}{[\beta - \alpha - 1]_{k-\alpha}[k - \alpha - 1]_{k-\alpha-1}} \\ &= -\frac{k - \alpha}{\beta - k} = \frac{\alpha - k}{\beta - k}. \end{aligned}$$

Similarly, if $\beta < \alpha$, either of the two term factors

$$g(k) = \begin{cases} \dfrac{1}{[\alpha - k]_{\alpha-\beta}} = [\beta - k]_{\beta-\alpha}, \\ [k - \alpha - 1]_{\beta-\alpha}, \end{cases}$$

gives the quotient as above and is defined for $k \leq \beta$ and for $\alpha < k$.

For the interval $(\beta, \alpha]$, we can choose the term factor

$$h(k) = (-1)^k \binom{\alpha - \beta - 1}{k - \beta - 1},$$

which has the quotient

$$q_h(k) = -\frac{\binom{\alpha-\beta-1}{k-\beta}}{\binom{\alpha-\beta-1}{k-\beta-1}} = \frac{\alpha - k}{\beta - k}$$

and is defined for $\beta < k \le \alpha$.

5.4 Hypergeometric Form

Sums of types I–II can be written as hypergeometric sums as well. To do that, it is only necessary to use (5.7) and the ascending factorial from (1.3) and (2.5), possibly (2.1). As an example consider the Gaussian hypergeometric function,

$${}_2F_1(a, b; c; z) = \sum_{k=0}^{\infty} \frac{(a)_k (b)_k}{(c)_k} \frac{z^k}{k!}.$$

With this we can write a canonical sum of type $\mathrm{II}(2, 2, z)$ as

$$\sum_{k=0}^{n} \binom{n}{k} [\alpha]_k [\beta]_{n-k} z^k = [\beta]_n \cdot {}_2F_1(-n, -\alpha; \beta - n + 1; z).$$

In general, the transformations are

$$\begin{aligned} &{}_pF_q(-n, a_1, \ldots; b_1, \ldots; z) \\ &\quad = \frac{(-1)^{nq}}{[-b_1]_n \cdots} \sum_{k=0}^{n} \binom{n}{k} [-a_1]_k \cdots [b_1 + n - 1]_{n-k} \cdots ((-1)^p z)^k \end{aligned}$$

and

$$\begin{aligned} &\sum_{k=0}^{n} \binom{n}{k} [\alpha_1]_k \cdots [\beta_1]_{n-k} \cdots z^k = \\ &\qquad [\beta_1]_n \cdots {}_pF_q(-n, -\alpha_1, \ldots; \beta_1 - n + 1, \ldots; (-1)^p z). \end{aligned}$$

5.5 The Classification Recipe

Given the formula (5.1), we first check whether the terms $t(c,n,k)$ are rational functions in k, in which case the sum is of type I. Next, if this is not the case, we extract factors of the form of harmonic numbers $H_{c,k}^{(m)}$ or the form (where we assume $0 \neq y \neq \pm d$)

$$[x+yk,d]_k,$$

to see if the types III–V can apply. Then we take the quotients of the rest of the terms, and if they are rational functions in k, we find the roots of the numerator and denominator. Now we know the possible type, II–V.

Next, we compare the roots with the limits to determine if the limits are natural, or if it is convenient to divide the sum as sums over several different intervals. This might be the case if there are several integral roots of the quotient in the original interval of summation.

Then we look in the table of contents under the heading $X(p,q,z)$ to see if the sum is known. If not, in the case of type II(p,q,z), we can try the Zeilberger algorithm, and in the case of type I, the Gosper algorithm.

5.6 Symmetric and Balanced Sums

The majority of known sums with several factors are what we will call either *symmetric* or *balanced.* The canonical form (5.8) is most convenient for recognizing these cases. But note that the binomial coefficient corresponds to a pair of factors by the transformation

$$\binom{n}{k} = \frac{[n]_k}{[k]_k} = \frac{[n]_k[n]_{n-k}}{[n]_n}.$$

So, in the canonical form we can have the arguments $a_1 = n, a_2, \ldots$ for k and $b_1 = n, b_2, \ldots$ for $n-k$.

Definition 5.5. A sum of type II$(p,p,\pm x)$ is called *symmetric* if we can write it in canonical form (5.8) with $a_j = b_j$, $j = 2, \ldots, p$.

This means that the canonical form of a symmetric sum becomes

$$S(c,n) = \sum_{k=0}^{n} \binom{n}{k} \prod_{j=2}^{p} [a_j]_k [a_j]_{n-k} x^k.$$

Remark 5.6. Symmetric sums of type $\mathrm{II}(p, p, (-1)^{p-1})$ share the property that the sums for n odd are zero, because of the change of sign when reversing the direction of summation.

Definition 5.7. A sum of type $\mathrm{II}(p, p, \pm x)$ is called *balanced* if we can write it in canonical form (5.8) with $a_j - b_j = 2a$, $j = 1, \ldots, p$ for some constant a.

Remark 5.8. In the case of balanced sums we must let $a_1 = n$ and $b_2 = n$ such that we get $b_1 = n - 2a$ and $a_2 = n + 2a$, while the rest of the arguments can be written as $a_j = c_j + a$ and $b_j = c_j - a$.

This means that the canonical form of a balanced sum becomes

$$S(c, n) = \sum_{k=0}^{n} \binom{n}{k} [n + 2a]_k [n - 2a]_{n-k} \prod_{j=3}^{p} [c_j + a]_k [c_j - a]_{n-k}\, x^k. \quad (5.10)$$

Definition 5.9. A sum is called *well-balanced* if it is balanced and can be written in canonical form (5.10) for some constant a, and such that for some $j > 2$ we further have $a_j + b_j = n - 2$.

If we have the canonical form and want to determine whether the sum is symmetric or balanced, the symmetric case is easy enough, and in the balanced case the roots of the quotient (5.5) must satisfy that the pairs have the constant sum $n - 1 + 2a$, so a can be found from the sum of all the roots divided by $2p$.

The well-balanced case means that for some pair α_j, β_j of corresponding roots their difference is 1, $\beta_j - \alpha_j = 1$, which gives the strange expressions $a_j = \alpha_j = \frac{n}{2} - 1 + a$, and therefore $\beta_j = \frac{n}{2} + a$ and $b_j = \frac{n}{2} - 1 - a$.

Remark 5.10. In the case of well-balanced sums, we can let $a_3 = \frac{n}{2} - 1 + a$ and $b_3 = \frac{n}{2} - 1 - a$. Hence, we can write the product

$$\begin{aligned} [a_3]_k [b_3]_{n-k} &= \left[\tfrac{n}{2} - 1 + a\right]_k \left[\tfrac{n}{2} - 1 - a\right]_{n-k} \\ &= \left[\tfrac{n}{2} - 1 + a\right]_k \left[\tfrac{n}{2} + a - k\right]_{n-k} (-1)^{n-k} \\ &= (-1)^n \tfrac{1}{2} \left[\tfrac{n}{2} - 1 + a\right]_{n-1} (n + 2a - 2k)\,(-1)^k. \end{aligned}$$

Dividing the constants out, we can get the special canonical form for well-balanced sums,

$$\begin{aligned} S(c, n) = \sum_{k=0}^{n} \binom{n}{k} [n + 2a]_k [n - 2a]_{n-k} &\Bigl(\prod_{j=3}^{p} [c_j + a]_k [c_j - a]_{n-k}\Bigr) \\ &\times (n + 2a - 2k)\,(-x)^k. \end{aligned}$$

Sometimes it is possible to find small deviations from these two basic conditions. If we have the condition fulfilled except for $a_1 = b_1 + p$ or

$a_1 - b_1 = 2a + p$ for some $p \in \mathbb{Z}$, called the *excess*, we designate these sums with the prefix *quasi-*. So we will talk about *quasi-symmetric* and *quasi-balanced* sums, etc.

5.7 Useful Transformations

In order to write a given sum in canonical form, it is often convenient to make one of the following transformations.

Lemma 5.11. *The products of constant length can be transformed to canonical form for a sum of limits* 0 *and* n *by*

$$[a-k]_p = [a-p]_k[n-1-a]_{n-k}(-1)^k \frac{(-1)^n}{[a-p]_{n-p}} \qquad a \notin \{p, p+1, \ldots, n-1\},$$

$$[a+k]_p = [-a-1]_k[n-p+a]_{n-k}(-1)^k \frac{(-1)^{n+p}}{[-a-1]_{n-p}} \qquad -a \notin \{1, 2, \ldots, n-p\}.$$

Proof: Trivial. □

Occasionally the following formulas are very useful:

$$[2n]_n = 2[2n-1]_n, \tag{5.11}$$

$$[2n]_n = 4^n \left[n-\tfrac{1}{2}\right]_n = (-4)^n \left[-\tfrac{1}{2}\right]_n, \tag{5.12}$$

$$\binom{2n}{n} = (-4)^n \binom{-\frac{1}{2}}{n}. \tag{5.13}$$

At other times, it is useful to have representations for evry second term, for example,

$$\binom{2n}{2k} = \frac{1}{\left[n-\frac{1}{2}\right]_n} \binom{n}{k} \left[n-\tfrac{1}{2}\right]_k \left[n-\tfrac{1}{2}\right]_{n-k}, \tag{5.14}$$

$$\binom{2n+1}{2k} = \frac{1}{\left[n-\frac{1}{2}\right]_n} \binom{n}{k} \left[n+\tfrac{1}{2}\right]_k \left[n-\tfrac{1}{2}\right]_{n-k}, \tag{5.15}$$

$$\binom{2n+1}{2k+1} = \frac{1}{\left[n-\frac{1}{2}\right]_n} \binom{n}{k} \left[n-\tfrac{1}{2}\right]_k \left[n+\tfrac{1}{2}\right]_{n-k}, \tag{5.16}$$

$$\binom{2n+2}{2k+1} = \frac{n+1}{\left[n+\frac{1}{2}\right]_{n+1}} \binom{n}{k} \left[n+\tfrac{1}{2}\right]_k \left[n+\tfrac{1}{2}\right]_{n-k}. \tag{5.17}$$

The formulas (5.14) and (5.15) can be combined and the formulas (5.16) and (5.17) combined in two ways as

$$\binom{n}{2k} = \frac{1}{\left[\lfloor\frac{n}{2}\rfloor - \frac{1}{2}\right]_{\lfloor\frac{n}{2}\rfloor}} \binom{\lfloor\frac{n}{2}\rfloor}{k} \left[\lceil\tfrac{n}{2}\rceil - \tfrac{1}{2}\right]_k \left[\lfloor\tfrac{n}{2}\rfloor - \tfrac{1}{2}\right]_{\lfloor\frac{n}{2}\rfloor - k},$$

$$\binom{n}{2k+1} = \frac{n}{2\left[\lfloor\frac{n-1}{2}\rfloor + \frac{1}{2}\right]_{\lfloor\frac{n-1}{2}\rfloor+1}} \binom{\lfloor\frac{n-1}{2}\rfloor}{k} \left[\lceil\tfrac{n-1}{2}\rceil - \tfrac{1}{2}\right]_k \times \left[\lfloor\tfrac{n-1}{2}\rfloor + \tfrac{1}{2}\right]_{\lfloor\frac{n-1}{2}\rfloor - k},$$

$$\binom{n+1}{2k+1} = \frac{n+1}{2\left[\lfloor\frac{n}{2}\rfloor + \frac{1}{2}\right]_{\lfloor\frac{n}{2}\rfloor+1}} \binom{\lfloor\frac{n}{2}\rfloor}{k} \left[\lceil\tfrac{n}{2}\rceil - \tfrac{1}{2}\right]_k \left[\lfloor\tfrac{n}{2}\rfloor + \tfrac{1}{2}\right]_{\lfloor\frac{n}{2}\rfloor - k}.$$

Another useful transformation is the following

$$\binom{n-k}{k} = \frac{(-1)^{\lfloor\frac{n}{2}\rfloor}}{[n]_{\lfloor\frac{n}{2}\rfloor}} \binom{\lfloor\frac{n}{2}\rfloor}{k} \left[\lceil\tfrac{n}{2}\rceil - \tfrac{1}{2}\right]_k \left[-\lceil\tfrac{n}{2}\rceil - 1\right]_{\lfloor\frac{n}{2}\rfloor - k} (-4)^k. \tag{5.18}$$

Or, if we divide into even and odd cases, we get

$$\binom{2n-k}{k} = \frac{(-1)^n}{[2n]_n} \binom{n}{k} \left[n - \tfrac{1}{2}\right]_k [-n-1]_{n-k} (-4)^k,$$

$$\binom{2n+1-k}{k} = \frac{(-1)^n}{[2n+1]_n} \binom{n}{k} \left[n + \tfrac{1}{2}\right]_k [-n-2]_{n-k} (-4)^k.$$

Furthermore, it is convenient to have computed the binomial of (5.18) times a factorial, for $m \in \mathbb{Z}$ and $m \vee 0 \leq k \leq \lfloor\frac{n}{2}\rfloor$:

$$[k]_m \binom{n-k}{k} = \frac{(-1)^{\lfloor\frac{n}{2}\rfloor - m}}{[n-2m]_{\lfloor\frac{n}{2}\rfloor - 2m}} \binom{\lfloor\frac{n}{2}\rfloor - m}{k-m} \times \left[\lceil\tfrac{n}{2}\rceil - m - \tfrac{1}{2}\right]_{k-m} \left[-\lceil\tfrac{n}{2}\rceil - 1\right]_{\lfloor\frac{n}{2}\rfloor - k} (-4)^{k-m}. \tag{5.19}$$

Or, if we decide to split into even and odd cases,

$$[k+m]_m \binom{2n+m-k}{k+m} = \frac{(-1)^n}{[2n]_{n-m}} \binom{n}{k} \times \left[n - \tfrac{1}{2}\right]_k [-n-m-1]_{n-k} (-4)^k, \tag{5.20}$$

$$[k+m]_m \binom{2n+m+1-k}{k+m} = \frac{(-1)^n}{[2n+1]_{n-m}} \binom{n}{k} \times \left[n + \tfrac{1}{2}\right]_k [-n-m-2]_{n-k} (-4)^k. \tag{5.21}$$

The similar formulas with $n+k$ instead of $n-k$ are much simpler and valid for $k, n \in \mathbb{N}$, $m \in \mathbb{Z}$, $-n \leq m$:

$$\binom{n+k}{k} = \binom{-n-1}{k}(-1)^k, \tag{5.22}$$

$$[k]_m \binom{n+k}{k} = [-n-1]_m \binom{-n-1-m}{k-m}(-1)^k. \tag{5.23}$$

5.8 Polynomial Factors

We can get rid of a polynomial factor in k by writing the polynomial as a sum of factorials (3.32) and then changing the expression using (2.8) and (2.2), as follows:

$$\binom{n}{k}[k]_m[a]_k[b]_{n-k} = [n]_m[a]_m\binom{n-m}{k-m}[a-m]_{k-m}[b]_{n-m-(k-m)}.$$

Chapter 6

Gosper's Algorithm

6.1 Gosper's Telescoping Algorithm

In 1978 R. William Gosper [Gosper 78] gave a beautiful algorithm for establishing possible indefinite sums of types I and II.

Assume we have a sum of the form,

$$T(a,k) = \sum t(a,k)\delta k, \tag{6.1}$$

with quotient function of the form

$$q_t(a,k) = \frac{(\alpha_1 - k)\cdots(\alpha_p - k)}{(\beta_1 - k)\cdots(\beta_q - k)}\chi. \tag{6.2}$$

If possible, we will choose the indices such that $\beta_i - \alpha_i \in \mathbb{N}$, $i = 1, \ldots, j$, and $\beta_i - \alpha_\ell \notin \mathbb{N}$, $i, \ell > j$. For each of the pairs with integral difference we will write the quotient, with $\alpha = \beta - m$, as

$$\frac{\alpha - k}{\beta - k} = \frac{\beta - m - k}{\beta - k} = \frac{(\beta - k - m)[\beta - k - 1]_{m-1}}{(\beta - k)[\beta - k - 1]_{m-1}} = \frac{[\beta - k - 1]_m}{[\beta - k]_m}. \tag{6.3}$$

Using (6.3) we can define the polynomial $f(k)$ of degree $m_1 + \cdots + m_j$ as

$$f(k) = [\beta_1 - k]_{m_1} \cdots [\beta_j - k]_{m_j}.$$

If we further define the polynomials of the rest of the factors,

$$g(k) = (\alpha_{j+1} - k)\cdots(\alpha_p - k)\chi,$$
$$h(k) = (\beta_{j+1} + 1 - k)\cdots(\beta_q + 1 - k),$$

then they satisfy that for no pair of roots do we have

$$\beta_i - \alpha_\ell \in \mathbb{N}, \tag{6.4}$$

and we can write the quotient (6.2) as

$$q_t(a,k) = \frac{f(k+1)}{f(k)} \frac{g(k)}{h(k+1)}. \tag{6.5}$$

With the quotient given in this form, we write the candidate for a solution to the sum (6.1) as

$$T(a,k) = \frac{h(k)s(k)t(a,k)}{f(k)}, \tag{6.6}$$

where $s(k)$ is an unknown function which we attempt to find.

If the function defined by (6.6) solves equation (6.1), then using equation (6.5) we must have

$$\begin{aligned} t(a,k) &= \Delta T(a,k) \\ &= \frac{h(k+1)s(k+1)t(a,k+1)}{f(k+1)} - \frac{h(k)s(k)t(a,k)}{f(k)} \\ &= \frac{s(k+1)t(a,k)g(k)}{f(k)} - \frac{h(k)s(k)t(a,k)}{f(k)} \\ &= t(a,k)\frac{s(k+1)g(k) - h(k)s(k)}{f(k)}. \end{aligned}$$

This means that if $s(k)$ solves the problem, it must satisfy the difference equation

$$f(k) = s(k+1)g(k) - s(k)h(k). \tag{6.7}$$

Now, it is obvious that the function $s(k)$ must be rational, i.e., we have polynomials P and Q without common roots, such that

$$s(k) = \frac{P(k)}{Q(k)}. \tag{6.8}$$

We want to prove that $s(k)$ is a polynomial, so we assume that Q is not constant. Then we can consider a non-root β, such that $\beta+1$ is a root of Q, and there exists a greatest integer, $N \in \mathbb{N}$, such that $\beta + N$ is a root of Q.

If we substitute (6.8) in (6.7) and multiply by $Q(k+1)Q(k)$, we get

$$f(k)Q(k+1)Q(k) = g(k)P(k+1)Q(k) - h(k)P(k)Q(k+1).$$

Now we apply this equation for $k = \beta$ and $k = \beta + N$ to obtain

$$\begin{aligned} 0 &= g(\beta)P(\beta+1)Q(\beta), \\ 0 &= h(\beta+N)P(\beta+N)Q(\beta+N+1). \end{aligned}$$

Neither P nor Q is zero at the chosen points, hence we have

$$0 = g(\beta),$$
$$0 = h(\beta + N),$$

contradicting the property of g and h that no pair of roots can have an integral difference of this sign. Hence s is a polynomial. Assume it looks like

$$s(k) = \alpha_d k^d + \alpha_{d-1} k^{d-1} + \cdots + \alpha_0, \quad \alpha_d \neq 0.$$

In order to solve (6.7) in s, we will estimate the size of d. We introduce the polynomials

$$G(k) = g(k) - h(k),$$
$$H(k) = g(k) + h(k).$$

Then we rewrite (6.7) as

$$2f(k) = G(k)(s(k+1) + s(k)) + H(k)(s(k+1) - s(k)) \qquad (6.9)$$

and observe that

$$s(k+1) + s(k) = 2\alpha_d k^d + \cdots \quad \text{of degree } d,$$
$$s(k+1) - s(k) = d\alpha_d k^{d-1} + \cdots \quad \text{of degree } d-1.$$

With deg meaning "degree" we then have four possibilities.

(1) If $\deg(G) = \deg(f)$, then s is a constant and H disappears from (6.9).

(2) If $\deg(G) \geq \deg(H)$, then we simply conclude that

$$d = \deg(f) - \deg(G). \qquad (6.10)$$

(3) If $d' = \deg(H) > \deg(G)$, then we consider

$$G(k) = \lambda' k^{d'-1} + \cdots,$$
$$H(k) = \lambda k^{d'} + \cdots \quad \lambda \neq 0.$$

Hence the right-hand side of (6.9) begins

$$(2\lambda' \alpha_d + \lambda d \alpha_d) k^{d+d'-1}.$$

This means that in general, if $2\lambda' + \lambda d \neq 0$, we have the formula

$$d = \deg(f) - \deg(H) + 1. \qquad (6.11)$$

(4) The last case is the exceptional one, where $\deg(H) > \deg(G)$ and $2\lambda' + \lambda d = 0$. But then the latter equation yields

$$d = -2\frac{\lambda'}{\lambda}. \tag{6.12}$$

This tells us that in the case of $\deg(H) > \deg(G)$ we can try one, two or three values for d, according to the integrability of the solution (6.12).

As soon as d is established, one can try to solve (6.7) in the $d+1$ coefficients.

6.2 An Example

Let us consider the following indefinite sum of type II(2, 2, 1) for a, b, c, and d any complex numbers,

$$T(a, b, c, d, k) = \sum \frac{[a]_k [b]_k}{[c-1]_k [d-1]_k} \delta k$$

with quotient

$$q(k) = \frac{(a-k)(b-k)}{(c-1-k)(d-1-k)}.$$

We will apply Gosper's algorithm to prove the following formula,

Theorem 6.1. *For any complex numbers, $a, b, c, d \in \mathbb{C}$ satisfying the condition that $p = a + b - c - d \in \mathbb{N}_0$, we have the indefinite summation formula*

$$\sum \frac{[a]_k [b]_k}{[c-1]_k [d-1]_k} \delta k = \frac{[a]_k [b]_k}{[c-1]_{k-1} [d-1]_{k-1}} \sum_{j=0}^{p} \frac{[p]_j [k-c-1]_j}{[a-c]_{j+1} [b-c]_{j+1}}. \tag{6.13}$$

Proof: As in general the differences (6.4) are not positive integers, we have immediately the polynomial $f(k) = 1$. The others are

$$g(k) = k^2 - (a+b)k + ab,$$
$$h(k) = k^2 - (c+d)k + cd.$$

Hence the sum and difference become

$$G(k) = (c + d - a - b)k + ab - cd,$$
$$H(k) = 2k^2 - (a + b + c + d)k + ab + cd.$$

So we are in case (2) or (3) with $\deg(G) < \deg(H)$, and after (6.11) we compute $\deg(f) - \deg(H) + 1 = 0 - 2 + 1 = -1$, leaving us with case (3) as the only possibility. Hence we have

$$\deg(s) = -2\frac{c+d-a-b}{2} = a+b-c-d.$$

So we have the restriction on the parameters that

$$p = a+b-c-d \in \mathbb{N}_0.$$

If $p = 0$, we have $s(k) = \alpha$ to be found by (6.7), i.e., from

$$1 = \alpha\,(g(k) - h(k)) = \alpha G(k) = \alpha(ab - cd).$$

So using $p = a+b-c-d = 0$, we have

$$\alpha = \frac{1}{ab-cd} = \frac{1}{(a-c)(b-c)}.$$

According to (6.6), we get the indefinite sum

$$\begin{aligned} T(a,b,c,d,k) &= \frac{(c-k)(d-k)}{(a-c)(b-c)} \cdot \frac{[a]_k[b]_k}{[c-1]_k[d-1]_k} \\ &= \frac{1}{(a-c)(b-c)} \cdot \frac{[a]_k[b]_k}{[c-1]_{k-1}[d-1]_{k-1}}. \end{aligned}$$

If $p = 1$, we have $s(k) = \alpha_1 k + \alpha_0$ to be found by (6.7), i.e., from

$$\begin{aligned} 1 &= (\alpha_1 k + \alpha_1 + \alpha_0)g(k) - (\alpha_1 k + \alpha_0)h(k) \\ &= (\alpha_1(ab - cd - a - b) - \alpha_0)k + \alpha_1 ab + \alpha_0(ab - cd), \end{aligned}$$

yielding two equations in (α_1, α_0),

$$\begin{aligned} 0 &= \alpha_1(ab - cd - a - b) - \alpha_0, \\ 1 &= \alpha_1 ab + \alpha_0(ab - cd), \end{aligned}$$

with the solutions

$$\begin{aligned} \alpha_0 &= \frac{ab - cd - a - b}{ab + (ab-cd)(ab-cd-a-b)}, \\ \alpha_1 &= \frac{1}{ab + (ab-cd)(ab-cd-a-b)}. \end{aligned}$$

To make it look nicer, we write the denominator as

$$
\begin{aligned}
ab+&(ab-cd)(ab-cd-a-b)\\
&= ab+(ab-cd)^2-(ab-cd)(a+b)\\
&= ab+(ab-cd)^2-(ab-cd)(1+c+d)\\
&= (ab-cd)^2-(ab-cd)(c+d)+cd\\
&= (ab-cd-c)(ab-cd-d)\\
&= (ab-c(a+b-c-1)-c)(ab-(c+1)(a+b-c-1)\\
&= (a-c)(b-c)(a-b-1)(b-c-1)\\
&= [a-c]_2[b-c]_2,
\end{aligned}
$$

and the numerator similarly as

$$ab-cd-a-b = ab-c(a+b-c-1)-a-b = (a-c-1)(b-c-1)-c-1.$$

According to (6.6) we get the indefinite sum

$$
\begin{aligned}
T(&a,b,c,d,k)\\
&= \frac{k+(a-c-1)(b-c-1)-c-1}{[a-c]_2[b-c]_2}\cdot\frac{[a]_k[b]_k}{[c-1]_{k-1}[d-1]_{k-1}}\\
&= \left(\frac{1}{(a-c)(b-c)}+\frac{k-c-1}{[a-c]_2[b-c]_2}\right)\cdot\frac{[a]_k[b]_k}{[c-1]_{k-1}[d-1]_{k-1}}.
\end{aligned}
$$

For each positive integral value of p, we can in principle continue this way, ending up with the formula for $p=a+b-c-d$,

$$\sum\frac{[a]_k[b]_k}{[c-1]_k[d-1]_k}\delta k = \frac{[a]_k[b]_k}{[c-1]_{k-1}[d-1]_{k-1}}\sum_{j=0}^{p}\frac{[p]_j[k-c-1]_j}{[a-c]_{j+1}[b-c]_{j+1}}. \tag{6.14}$$

But it is easier to proceed with guessing (6.14) and then proving it by induction. □

Corollary 6.2. *For any complex numbers, $a,b,c,d\in\mathbb{C}$ satisfying the condition that $a+b-c-d=0$, we have the indefinite summation formula*

$$\sum\frac{[a]_k[b]_k}{[c-1]_k[d-1]_k}\delta k = \frac{1}{(a-c)(b-c)}\cdot\frac{[a]_k[b]_k}{[c-1]_{k-1}[d-1]_{k-1}}. \tag{6.15}$$

Proof: Obvious. □

Chapter 7

Sums of Type II(1,1,z)

7.1 The Binomial Theorem

The most important, famous and oldest formula of type II$(1, 1, z)$ is the *binomial theorem*,

$$\sum_{k=0}^{n} \binom{n}{k} x^k y^{n-k} = (x+y)^n, \tag{7.1}$$

discovered around the year 1000 by Al-Karajī in Baghdad [Edwards 87]. It is of type II$(1, 1, z)$ with $z = \frac{x}{y}$. (For proof see Theorem 8.1.)

Remark 7.1. The binomial formula generalizes to $n \in \mathbb{C}$ in the form of an infinite series, the Taylor series, applying the general binomial coefficients, defined by (1.5)

$$\sum_{k=0}^{\infty} \binom{n}{k} x^k y^{n-k} = (x+y)^n \text{ for } |x| < |y|. \tag{7.2}$$

The formula (7.1) is of type II$(1, 1, z)$ with natural limits and quotient

$$q_{\text{binom}}((x, y), n, k) = \frac{n-k}{-1-k} \left(-\frac{x}{y}\right).$$

An indefinite formula of type II$(1, 1, 1)$ is

$$\sum \frac{[\alpha]_k}{[\beta]_k} \delta k = \frac{1}{\alpha - \beta - 1} \frac{[\alpha]_k}{[\beta]_{k-1}}, \tag{7.3}$$

with quotient

$$q(\alpha, \beta, k) = \frac{\alpha - k}{\beta - k}.$$

Proof:

$$\Delta\frac{[\alpha]_k}{[\beta]_{k-1}} = \frac{[\alpha]_{k+1}}{[\beta]_k} - \frac{[\alpha]_k}{[\beta]_{k-1}} = \frac{[\alpha]_k(\alpha - k - \beta + k - 1)}{[\beta]_k}.$$

□

We have a formula of type II$(1,1,-1)$ due to Tor B. Staver [Staver 47]:

$$\sum_{k=0}^{n} \frac{1}{\binom{n}{k}} = (n+1)2^{-n-1}\sum_{k=1}^{n+1}\frac{2^k}{k}, \tag{7.4}$$

with quotient

$$q_t(k) = \frac{-1-k}{n-k}(-1).$$

Proof of (7.4), *Staver's formula:* Define

$$S(n,p) = \sum_{k=0}^{n} \frac{k^p}{\binom{n}{k}}.$$

Then we have

$$S(n,0) = \frac{1}{n}\left(\sum_{k=0}^{n}\frac{n-k}{\binom{n}{k}} + \sum_{k=0}^{n}\frac{k}{\binom{n}{k}}\right) = \frac{2}{n}S(n,1). \tag{7.5}$$

On the other hand, we also have

$$\begin{aligned} S(n,0) = 1 + \sum_{k=1}^{n}\frac{k}{n\binom{n-1}{k-1}} &= 1 + \frac{1}{n}\sum_{k=0}^{n-1}\frac{k}{\binom{n-1}{k}} + \frac{1}{n}\sum_{k=0}^{n-1}\frac{1}{\binom{n-1}{k}} \\ &= 1 + \frac{1}{n}S(n-1,1) + \frac{1}{n}S(n-1,0). \end{aligned}$$

When we apply (7.5) for $n-1$, we get

$$S(n,0) = 1 + \frac{1}{n}\frac{n-1}{2}S(n-1,0) + \frac{1}{n}S(n-1,0) = \frac{n+1}{2n}S(n-1,0) + 1.$$

The homogeneous equation has the solution $\frac{n+1}{2^n}$, so we are looking for a solution of the form $\phi(n)\frac{n+1}{2^n}$, i.e., we look at the equation

$$\phi(n)\frac{n+1}{2^n} = \phi(n-1)\frac{n+1}{2n}\frac{n}{2^{n-1}} + 1,$$

or

$$\phi(n) - \phi(n-1) = \frac{2^n}{n+1},$$

which solves the problem in the form

$$S(n,0) = \frac{n+1}{2^n}\sum_{k=0}^{n}\frac{2^k}{k+1} = (n+1)2^{-n-1}\sum_{k=1}^{n+1}\frac{2^k}{k}.$$

□

7.2 Gregory Galperin and Hillel Gauchman's Problem

In 2004 Gregory Galperin and Hillel Gauchman posed the problem of proving the following identity [Galperin and Gauchman 04]:

$$\sum_{k=1}^{n} \frac{1}{k\binom{n}{k}} = \frac{1}{2^{n-1}} \sum_{k \text{ odd}}^{n} \frac{\binom{n}{k}}{k}. \tag{7.6}$$

By (7.4) the left side is

$$\sum_{k=1}^{n} \frac{1}{k\binom{n}{k}} = \frac{1}{n}\sum_{k=1}^{n} \frac{1}{\binom{n-1}{k-1}} = 2^{-n}\sum_{k=1}^{n} \frac{2^k}{k} = \frac{1}{2^{n-1}}\sum_{k=1}^{n} \frac{2^{k-1}}{k}. \tag{7.7}$$

Now consider the sum on the right side of (7.6). The difference in n is

$$\sum_{k \text{ odd}} \frac{\binom{n}{k}}{k} - \sum_{k \text{ odd}} \frac{\binom{n-1}{k}}{k} = \sum_{k \text{ odd}} \frac{\binom{n-1}{k-1}}{k} = \frac{1}{n}\sum_{k \text{ odd}} \binom{n}{k}.$$

As we have

$$\sum_{k \text{ odd}} \binom{n}{k} + \sum_{k \text{ even}} \binom{n}{k} = \sum_{k} \binom{n}{k} = 2^n,$$

$$\sum_{k \text{ even}} \binom{n}{k} - \sum_{k \text{ odd}} \binom{n}{k} = \sum_{k} \binom{n}{k}(-1)^k = 0,$$

we get the desired result from (7.7)

$$\sum_{k \text{ odd}} \frac{\binom{n}{k}}{k} = \sum_{k=1}^{n} \frac{2^{k-1}}{k}.$$

Chapter 8

Sums of Type II(2,2,z)

8.1 The Chu–Vandermonde Convolution

The most important sum of type II(2,2,z) is the generalization of one of the oldest formulas, *the Chu–Vandermonde convolution,*

$$\sum_{k=0}^{n} \binom{x}{k} \binom{y}{n-k} = \binom{x+y}{n}, \tag{8.1}$$

with quotient

$$q_{CV} = \frac{(n-k)(x-k)}{(-1-k)(n-1-y-k)}.$$

It appeared in 1303 in the famous Chinese algebra text *Precious Mirror of the Four Elements,*[1] by Chu Shih-Chieh [Chu 03, Hoe 77]. The formula was rediscovered in 1772 by A. T. Vandermonde [Vandermonde 72]. The combinatorial interpretation is to consider an urn with x red and y blue balls. Now we can take n balls in $n+1$ ways, k red and $n-k$ blue. Together these numbers add up to the number of ways to take n out of $x+y$ balls while ignoring the color. The generalization we consider here is to allow $x, y \in \mathbb{C}$ and a general step size d, which for $d = 0$ includes the binomial theorem of type II$(1, 1, z)$, (7.1)

Rather than consider the expression (8.1) with generalized binomial coefficients, it is better to multiply the equation by the integral denominator $[n]_n$. Then (8.1) can be written as

$$\sum_{k=0}^{n} \binom{n}{k} [x]_k [y]_{n-k} = [x+y]_n. \tag{8.2}$$

[1]This is the usual translation of the title, which might better be: *Textbook on Four Variables.* Chu uses the four directions, north, east, south, west as variables or unknowns.

But this expression allows a generalization with arbitrary step size:

Theorem 8.1. *For $n \in \mathbb{N}_0$, $x, y \in \mathbb{C}$, and $d \in \mathbb{C}$, we have*

$$\sum_{k=0}^{n} \binom{n}{k} [x,d]_k [y,d]_{n-k} = [x+y,d]_n. \tag{8.3}$$

Proof: Let

$$S(n) := \sum_{k} \binom{n}{k} [x,d]_k [y,d]_{n-k}.$$

Using (2.7) we replace $\binom{n}{k}$ by $\binom{n-1}{k-1}+\binom{n-1}{k}$ and split the sum into two sums. In the sum with $\binom{n-1}{k-1}$ we substitute $k+1$ for k. Corresponding terms in the two sums now have $\binom{n-1}{k}[x,d]_k[y,d]_{n-k-1}$ as a common factor. Using this we obtain

$$\sum_{k} \binom{n-1}{k} [x,d]_k [y,d]_{n-k-1} \big(x - kd + y - (n-k-1)d\big)$$
$$= S(n-1)\big(x+y-(n-1)d\big).$$

Recursion now yields

$$S(n) = S(n-k)[x+y-(n-k)d, d]_k.$$

Since $S(0)=1$, we obtain (8.3). □

In the case of $d \neq 0$, the quotient becomes

$$q_t(a,b,d,n,k) = \frac{(n-k)(\frac{a}{d}-k)}{(-1-k)(n-1-\frac{b}{d}-k)},$$

which means that we can always compare this to a formula with $d = 1$, provided we have $d \neq 0$. Hence, if we come across a formula

$$T(a,b,n) = \sum_{k=0}^{n} t(a,b,n,k)$$

with quotient

$$q_t(a,b,n,k) = \frac{t(a,b,n,k+1)}{t(a,b,n,k)} = \frac{(n-k)(a-k)}{(-1-k)(b-k)},$$

then the formula has type II(2, 2, 1) and is equivalent to Chu–Vandermonde (8.3). Hence the sum must be

$$T(a,b,n) = \frac{t(a,b,n,0)}{[n-1-b]_n}[a+n-1-b]_n. \tag{8.4}$$

As an example, consider the formulas

$$\sum_{k=0}^{n}\binom{n}{k}(-1)^{n-k}[c+kd,d]_m=[c,d]_{m-n}d^n[m]_n, \tag{8.5}$$

$$\sum_{k=0}^{n}\binom{n}{k}(-1)^{k}[c-kd,d]_m=[c-nd,d]_{m-n}d^n[m]_n. \tag{8.6}$$

The quotient in (8.5) is

$$-\frac{\binom{n}{k+1}}{\binom{n}{k}}\frac{[c+(k+1)d,d]_m}{[c+kd,d]_m}=\frac{n-k}{-1-k}\cdot\frac{-\frac{c}{d}-1-k}{m-\frac{c}{d}-1-k}.$$

Hence we get from (8.4) that the sum must be

$$\frac{(-1)^n[c,d]_m}{\left[n-1-m+1+\frac{c}{d}\right]_n}\left[-\frac{c}{d}-1+n-1-m+1+\frac{c}{d}\right]_n=[c,d]_{m-n}d^n[m]_n, \tag{8.7}$$

where we have used formulas (2.1), (2.2), (2.3), and (2.5). Formula (8.6) follows from (8.5) by reversing the direction of summation.

With $d=1$ the formulas (8.5)–(8.6) are very frequently used in the forms

$$\sum_{k=0}^{n}\binom{n}{k}(-1)^{n-k}[c+k]_m=[c]_{m-n}[m]_n, \tag{8.8}$$

$$\sum_{k=0}^{n}\binom{n}{k}(-1)^{k}[c-k]_m=[c-n]_{m-n}[m]_n. \tag{8.9}$$

Remark 8.2. For $0\le m<n$ the right side becomes 0.

Remark 8.3. For $c=0$ the right side becomes 0, except for $m=n$. Hence

$$\sum_{k=0}^{n}\binom{n}{k}(-1)^{n-k}[k]_m=\delta_{mn}[n]_n=\delta_{mn}[m]_m.$$

Remark 8.4. As we can write

$$k^m=\sum_{j=1}^{m}\mathfrak{S}_m^{(j)}[k]_j,$$

where $\mathfrak{S}_m^{(j)}$ are the Stirling numbers of the second kind as defined in (3.32), we get the general formula

$$\sum_{k=0}^{n}\binom{n}{k}(-1)^{n-k}k^m=\begin{cases}0 & \text{for } 0\le m<n,\\ \mathfrak{S}_m^{(n)}[n]_n & \text{for } n\le m.\end{cases} \tag{8.10}$$

Remark 8.5. For $m = -1$ we can apply (2.3) (with c replaced by $c - d$) to get

$$\sum_{k=0}^{n} \binom{n}{k} (-1)^{n-k} \frac{1}{c+kd} = \frac{d^n[-1]_n}{[c+nd,d]_{n+1}} = \frac{d^n(-1)^n[n]_n}{[c+nd,d]_{n+1}}$$

or

$$\frac{1}{c(c+d)\cdots(c+nd)} = \frac{1}{d^n[n]_n} \sum_{k=0}^{n} \frac{(-1)^k\binom{n}{k}}{c+kd},$$

i.e., the partial fraction for a polynomial with equidistant roots. For $d = 1$ we even get

$$\frac{1}{(c)_{n+1}} = \frac{1}{[c+n]_{n+1}} = \frac{1}{[n]_n} \sum_{k=0}^{n} \frac{(-1)^k\binom{n}{k}}{c+k}, \tag{8.11}$$

i.e., the partial fraction for a polynomial with consecutive integral roots.

8.2 A Simple Example

Consider the formula

$$\sum_{k=0}^{n} (-1)^k \binom{n-k}{k} \binom{n-2k}{m-k} = 1 \quad m \le n. \tag{8.12}$$

In order to analyze such a formula we may find the quotients. We find

$$\begin{aligned} q(k) &= -\frac{[n-k-1]_{k+1}}{[n-k]_k} \cdot \frac{[k]_k}{[k+1]_{k+1}} \cdot \frac{[n-2k-2]_{m-k-1}}{[n-2k]_{m-k}} \\ &\quad \times \frac{[m-k]_{m-k}}{[m-k-1]_{m-k-1}} \\ &= -\frac{[n-2k]_2}{n-k} \cdot \frac{1}{k+1} \cdot \frac{n-m-k}{[n-2k]_2} \cdot \frac{m-k}{1} = \frac{(m-k)(n-m-k)}{(-1-k)(n-k)}. \end{aligned}$$

This is a quotient of type II(2, 2, 1), i.e., of the Chu–Vandermonde convolution (8.2), if the limits are right. Now it is obvious that terms for $k > m$ vanish, so we can compute the value of (8.12) as

$$\begin{aligned} &\sum_{k=0}^{m} (-1)^k \binom{n-k}{k} \binom{n-2k}{m-k} \\ &\qquad = \frac{\binom{n}{m}}{[m-n-1]_m} \sum_{k=0}^{m} \binom{m}{k} [n-m]_k [m-n-1]_{m-k} \\ &\qquad = \frac{[n]_m[-1]_m}{[m]_m[m-n-1]_m} = 1. \end{aligned} \tag{8.13}$$

After realizing that the sum is just a plain Chu–Vandermonde sum, one can look for a shortcut to this formula. First, we assume $m \le \frac{n}{2}$, as otherwise the terms vanish for $k > n - m$. Then we multiply and divide by $m!$ to write

$$\begin{aligned}\sum_{k=0}^{m}(-1)^k\binom{n-k}{k}\binom{n-2k}{m-k} &= \frac{1}{m!}\sum_{k=0}^{m}(-1)^k\binom{m}{k}[n-k]_k[n-2k]_{m-k}\\ &= \frac{1}{m!}\sum_{k=0}^{m}\binom{m}{k}[n-k]_m(-1)^k\\ &= \frac{1}{m!}[n-m]_{m-m}[m]_m = 1, \qquad (8.14)\end{aligned}$$

according to (8.9).

8.3 The Laguerre Polynomials

When we learn some clever formula like the expressions for the Laguerre polynomials

$$L_n(x) = e^x\frac{d^n}{dx^n}(x^n e^{-x}) = n!\sum_{j=0}^{n}\binom{n}{j}\frac{(-x)^j}{j!},$$

none of the equalities are immediately verified. The proof of the second equality proceeds as follows:

$$\begin{aligned}e^x\frac{d^n}{dx^n}(x^n e^{-x}) &= \left(\sum_{j=0}^{\infty}\frac{x^j}{j!}\right)\frac{d^n}{dx^n}\left(x^n\sum_{k=0}^{\infty}\frac{(-x)^k}{k!}\right)\\ &=\left(\sum_{j=0}^{\infty}\frac{x^j}{j!}\right)\left(\sum_{k=0}^{\infty}\frac{(-1)^k[n+k]_n x^k}{k!}\right) = \sum_{k=0}^{\infty}x^k\sum_{j=0}^{k}\frac{(-1)^j[n+j]_n}{j!(k-j)!}\\ &=\sum_{k=0}^{\infty}\frac{x^k}{k!}\sum_{j=0}^{k}\binom{k}{j}(-1)^j[n+j]_n = \sum_{k=0}^{\infty}\frac{x^k}{k!}(-1)^k[n]_k[n]_{n-k}\\ &=n!\sum_{k=0}^{n}\binom{n}{k}\frac{(-x)^k}{k!}.\end{aligned}$$

The crucial point was, of course, the step

$$\sum_{j=0}^{k}\binom{k}{j}(-1)^j[n+j]_n = (-1)^k[n]_k[n]_{n-k},$$

which is just (8.9).

The systematic way of recognizing the relevant form is to consider the quotient of two consecutive terms as a rational function of j,

$$q(j) = \frac{(k-j)(-k-1-j)}{(-1-j)(-1-j)},$$

and then recognize this as the special case with $m = 0$, $c = p = k$, of

$$q(j) = \frac{(k-j)(m-c-1-j)}{(m-1-j)(m-1+p-c-j)}, \tag{8.15}$$

the quotient of the Chu–Vandermonde (8.2), translated to

$$\sum_{j=m}^{k} \binom{k-m}{j-m} (-1)^{k-j} [c-m+j]_p = [p]_{k-m} [c]_{p-k+m}. \tag{8.16}$$

8.4 Moriarty's Formulas

The so-called Moriarty formulas are as given in H. W. Gould's table [Gould 72b, (3.177),(3.178)]:

$$\sum_{k=0}^{n-p} \binom{2n+1}{2p+2k+1} \binom{p+k}{k} = \binom{2n-p}{p} 2^{2n-2p}, \tag{8.17}$$

$$\sum_{k=0}^{n-p} \binom{2n+1}{2p+2k} \binom{p+k}{k} = \frac{n}{2n-p} \binom{2n-p}{p} 2^{2n-2p}. \tag{8.18}$$

Because of their ugliness, these formulas were named after Professor Moriarty by H. T. Davis in 1962, [Davis 62], with reference to the short story by A. Conan Doyle, *The Adventure of the Final Problem.* Doyle writes about the professor: "At the age of twenty-one he wrote a treatise upon the Binomial Theorem which had a European vogue. On the strength of it, he won the Mathematical Chair at one of our smaller Universities."

As pointed out by Gould [Gould 72a, Gould 74, Gould 76], Moriarty was a master of disguise, and these formulas are nothing but disguises of the formulas for the coefficients to the Chebysheff polynomials. Furthermore, eventually they prove to be the Chu–Vandermonde formula (8.2) in disguise.

The relation to the Chebysheff polynomials comes straightforwardly from the binomial theorem (7.1). Let $c = \cos x$ and $s = \sin x$; then we compute the real and imaginary parts of

$$
\begin{aligned}
(c+is)^n &= \sum_{k=0}^{n} \binom{n}{k} c^{n-k} (is)^k \\
&= \sum_k \binom{n}{2k} c^{n-2k} \left(-s^2\right)^k + is \sum_k \binom{n}{2k+1} c^{n-1-2k} \left(-s^2\right)^k \\
&= \sum_k \binom{n}{2k} c^{n-2k} \left(c^2-1\right)^k + is \sum_k \binom{n}{2k+1} c^{n-1-2k} \left(c^2-1\right)^k \\
&= \sum_k \binom{n}{2k} c^{n-2k} \sum_j \binom{k}{j} c^{2(k-j)} (-1)^j \\
&\quad + is \sum_k \binom{n}{2k+1} c^{n-1-2k} \sum_j \binom{k}{j} c^{2(k-j)} (-1)^j \\
&= \sum_j (-1)^j c^{n-2j} \sum_k \binom{n}{2k} \binom{k}{j} \\
&\quad + is \sum_j (-1)^j c^{n-1-2j} \sum_k \binom{n}{2k+1} \binom{k}{j}. \qquad (8.19)
\end{aligned}
$$

Rectifying the limits of summation yields the left sides of the formulas (8.17) and (8.18).

Rather than working on these cumbersome expressions, we just compute the quotients. For (8.18) it is

$$
\begin{aligned}
q(k) &= \frac{(2n-2p-2k)(2n-2p-2k+1)(p+k+1)}{(2p+2k+2)(2p+2k+1)(k+1)} \\
&= \frac{(n-p-k)\left(n-p+\frac{1}{2}-k\right)}{(-1-k)\left(-p-\frac{1}{2}-k\right)}. \qquad (8.20)
\end{aligned}
$$

Hence it is a Chu–Vandermonde sum with limit $n-p$ and arguments $n-p+\frac{1}{2}$ and $n-\frac{1}{2}$. So the sum must be equal to

$$
\binom{2n+1}{2p} \frac{[2n-p]_{n-p}}{\left[n-\frac{1}{2}\right]_{n-p}} = \frac{2n+1}{2n+1-2p} \binom{2n-p}{p} 4^{n-p}.
$$

The quotient of (8.17) is

$$
\begin{aligned}
q(k) &= \frac{(2n-2p-2k)(2n-2p-2k-1)(p+k+1)}{(2p+2k+2)(2p+2k+3)(k+1)} \\
&= \frac{(n-p-k)\left(n-p-\frac{1}{2}-k\right)}{(-1-k)\left(-p-\frac{3}{2}-k\right)}.
\end{aligned}
$$

Hence it is a Chu–Vandermonde sum with limit $n-p$ and arguments $n-p-\frac{1}{2}$ and $n+\frac{1}{2}$. So the sum must be equal to

$$\binom{2n+1}{2p+1}\frac{[2n-p]_{n-p}}{\left[n+\frac{1}{2}\right]_{n-p}} = \binom{2n-p}{p}4^{n-p}.$$

8.5 An Example of Matrices as Arguments

The solution to a *difference equation* is essentially the computation of the powers $\mathbf{A}^n$ ($n \in \mathbb{N}$) of a matrix

$$\mathbf{A} = \begin{pmatrix} a & b \\ c & d \end{pmatrix}.$$

We will need the half trace and the determinant of the matrix, and the discriminant for the characteristic equation too:

$$\Theta = \frac{a+d}{2}, \tag{8.21}$$

$$D = ad - bc, \tag{8.22}$$

$$\Delta = \Theta^2 - D. \tag{8.23}$$

With these notations we can write the Cayley–Hamilton formula (4.30) for $\mathbf{A}$ as

$$\mathbf{A}^2 - 2\Theta\mathbf{A} + D\,\mathbf{I} = \mathbf{O},$$

or more conveniently for us,

$$(\mathbf{A} - \Theta\,\mathbf{I})^2 = \Delta\,\mathbf{I}.$$

For this reason we will find it useful to write

$$\mathbf{A} = \Theta\,\mathbf{I} + (\mathbf{A} - \Theta\,\mathbf{I}). \tag{8.24}$$

Because $\mathbf{I}$ commutes with any matrix, we can apply the binomial formula (7.1) to the sum (8.24).

$$\begin{aligned}
\mathbf{A}^n &= (\Theta\,\mathbf{I} + (\mathbf{A} - \Theta\,\mathbf{I}))^n \\
&= \sum_k \binom{n}{2k}\Theta^{n-2k}(\mathbf{A} - \Theta\,\mathbf{I})^{2k} + \sum_k \binom{n}{2k+1}\Theta^{n-2k-1}(\mathbf{A} - \Theta\,\mathbf{I})^{2k+1} \\
&= \sum_k \binom{n}{2k}\Theta^{n-2k}\Delta^k\,\mathbf{I} + \sum_k \binom{n}{2k+1}\Theta^{n-1-2k}\Delta^k(\mathbf{A} - \Theta\,\mathbf{I}).
\end{aligned} \tag{8.25}$$

In the case of $\Delta = 0$ we only get the terms for $k = 0$, i.e.,

$$\mathbf{A}^n = \Theta^n \mathbf{I} + n\Theta^{n-1}(\mathbf{A} - \Theta\,\mathbf{I}) = \Theta^n \left(\mathbf{I} + \frac{n}{\Theta}(\mathbf{A} - \Theta\,\mathbf{I})\right).$$

If $D = 0$, we have $\Delta = \Theta^2$ and hence the whole sum reduces to

$$\begin{aligned}\mathbf{A}^n &= \sum_k \binom{n}{2k} \Theta^n \mathbf{I} + \sum_k \binom{n}{2k+1} \Theta^{n-1}(\mathbf{A} - \Theta\,\mathbf{I})\\ &= 2^{n-1}\Theta^n \mathbf{I} + 2^{n-1}\Theta^{n-1}(\mathbf{A} - \Theta\,\mathbf{I})\\ &= (2\Theta)^{n-1}\mathbf{A}.\end{aligned}$$

In the case of $\Delta > 0$ we remark that

$$\left(\Theta \pm \sqrt{\Delta}\right)^n = \sum_k \binom{n}{2k} \Theta^{n-2k}\Delta^k \pm \sqrt{\Delta} \sum_k \binom{n}{2k+1} \Theta^{n-1-2k}\Delta^k.$$

Solving these two equations, we get

$$\begin{aligned}\sum_k \binom{n}{2k} \Theta^{n-2k}\Delta^k &= \frac{\left(\Theta + \sqrt{\Delta}\right)^n + \left(\Theta - \sqrt{\Delta}\right)^n}{2},\\ \sum_k \binom{n}{2k+1} \Theta^{n-1-2k}\Delta^k &= \frac{\left(\Theta + \sqrt{\Delta}\right)^n - \left(\Theta - \sqrt{\Delta}\right)^n}{2\sqrt{\Delta}}.\end{aligned}$$

So we get the formula

$$\begin{aligned}\mathbf{A}^n = &\frac{\left(\Theta + \sqrt{\Delta}\right)^n + \left(\Theta - \sqrt{\Delta}\right)^n}{2}\mathbf{I}\\ &+ \frac{\left(\Theta + \sqrt{\Delta}\right)^n - \left(\Theta - \sqrt{\Delta}\right)^n}{2\sqrt{\Delta}}(\mathbf{A} - \Theta\,\mathbf{I}).\end{aligned}$$

In the case of $\Delta < 0$, we remark that from (8.23) we have $\Delta = \Theta^2 - D$, and hence

$$\Delta^k = \left(\Theta^2 - D\right)^k = \sum_j \binom{k}{j} \Theta^{2k-2j}(-D)^j. \qquad (8.26)$$

Substituting (8.26) in (8.25) allows us to proceed as follows:

$$\begin{aligned}
\mathbf{A}^n &= \sum_k \binom{n}{2k} \Theta^{n-2k} \sum_j \binom{k}{j} \Theta^{2k-2j} (-D)^j \mathbf{I} \\
&\quad + \sum_k \binom{n}{2k+1} \Theta^{n-1-2k} \sum_j \binom{k}{j} \Theta^{2k-2j} (-D)^j (\mathbf{A} - \Theta \mathbf{I}) \\
&= \sum_j \Theta^{n-2j} (-D)^j \sum_k \binom{n}{2k} \binom{k}{j} \mathbf{I} \\
&\quad + \sum_j \Theta^{n-1-2j} (-D)^j \sum_k \binom{n}{2k+1} \binom{k}{j} (\mathbf{A} - \Theta \mathbf{I}) \\
&= \sum_j \Theta^{n-2j} (-D)^j \frac{n}{2} \binom{n-j}{j} \frac{2^{n-2j}}{n-j} \mathbf{I} \\
&\quad + \sum_j \Theta^{n-1-2j} (-D)^j \binom{n-1-j}{j} 2^{n-1-2j} (\mathbf{A} - \Theta \mathbf{I}).
\end{aligned}$$

In the case of $\Delta < 0$ we always have $D > 0$, hence we can proceed,

$$\begin{aligned}
\mathbf{A}^n &= \left(\sqrt{D}\right)^n \frac{n}{2} \sum_j \binom{n-j}{j} \frac{(-1)^j}{n-j} \left(\frac{2\Theta}{\sqrt{D}}\right)^{n-2j} \mathbf{I} \\
&\quad + \left(\sqrt{D}\right)^{n-1} \sum_j \binom{n-1-j}{j} (-1)^j \left(\frac{2\Theta}{\sqrt{D}}\right)^{n-1-2j} (\mathbf{A} - \Theta \mathbf{I}).
\end{aligned}$$

The sums can be recognized as the Chebysheff polynomials (8.19), so using the rewritten

$$\sin\left(\arccos\left(\frac{\Theta}{\sqrt{D}}\right)\right) = \sqrt{1 - \left(\frac{\Theta}{\sqrt{D}}\right)^2} = \sqrt{\frac{D - \Theta^2}{D}} = \sqrt{\frac{-\Delta}{D}},$$

we proceed to

$$\begin{aligned}
\mathbf{A}^n &= (\sqrt{D})^n \left(T_n\left(\frac{\Theta}{\sqrt{D}}\right) \mathbf{I} + \frac{1}{\sqrt{D}} U_{n-1}\left(\frac{\Theta}{\sqrt{D}}\right) (\mathbf{A} - \Theta \mathbf{I}) \right) \\
&= (\sqrt{D})^n \left(\cos\left(n \arccos\left(\frac{\Theta}{\sqrt{D}}\right)\right) \mathbf{I} + \frac{\sin\left(n \arccos\left(\frac{\Theta}{\sqrt{D}}\right)\right)}{\sqrt{-\Delta}} (\mathbf{A} - \Theta \mathbf{I}) \right).
\end{aligned}$$

8.6 Joseph M. Santmyer's Problem

In 1994 Joseph M. Santmyer [Santmyer 94] posed the following problem.

If M, N are integers satisfying $1 \leq m \leq n-1$, prove that

$$\binom{2n-m-1}{2n-2m-1} - \binom{n-1}{m} = \sum_k \sum_j \binom{k+j}{k} \binom{2n-m-2k-j-3}{2(n-m-k-1)}.$$

Solution. We have nonzero terms only for $0 \leq k \leq n-m-1$ and $0 \leq j \leq m-1$, hence the sum is finite.

We write the sum as

$$\begin{aligned} S &= \sum_{k=0}^{n-m-1} \sum_{j=0}^{m-1} \frac{[k+j]_k}{[k]_k} \cdot \frac{[2n-m-2k-j-3]_{m-1-j}}{[m-1-j]_{m-1-j}} \\ &= \sum_{k=0}^{n-m-1} \frac{1}{[k]_k} \sum_{j=0}^{m-1} [k+j]_k \frac{[2n-m-2k-3]_{m-1}[m-1]_j}{[2n-m-2k-3]_j[m-1]_{m-1}} \\ &= \frac{1}{[m-1]_{m-1}} \sum_{k=0}^{n-m-1} \frac{[2n-m-2k-3]_{m-1}}{[k]_k} \sum_{j=0}^{m-1} \frac{[k+j]_k[m-1]_j}{[2n-m-2k-3]_j}. \end{aligned}$$

In this form the inner sum is recognizable. We need to compute the quotient of two consecutive terms:

$$q(j) = \frac{(m-1-j)(-k-1-j)}{(-1-j)(2n-m-2k-3-j)}.$$

This is the quotient of the Chu–Vandermonde formula with parameters $-k-1$ and $2m-2n+2k+1$. Hence the sum is easily computed as

$$\frac{[k]_k[2m-2n+k]_{m-1}}{[2m-2n+2k+1]_{m-1}} = \frac{[k]_k[2n-2m-k+m-2]_{m-1}}{[2n-2m-2k-1+m-2]_{m-1}}.$$

Substitution of this result in the sum gives

$$\begin{aligned} S &= \frac{1}{[m-1]_{m-1}} \sum_{k=0}^{n-m-1} \frac{[2n-m-2k-3]_{m-1}[k]_k[2n-m-2-k]_{m-1}}{[k]_k[2n-m-2k-3]_{m-1}} \\ &= \frac{1}{[m-1]_{m-1}} \sum_{k=0}^{n-m-1} [2n-m-2-k]_{m-1}. \end{aligned}$$

This sum is easily calculated, but in order to get the desired result we reverse the order of summation,

$$S = \frac{1}{[m-1]_{m-1}} \sum_{k=0}^{n-m-1} [n-1+k]_{m-1}.$$

The quotient of this sum becomes

$$q(k) = \frac{n+k}{n-m+1+k} = \frac{-n-k}{-n+m-1-k},$$

so that the sum is equivalent to (7.3) with $a = -n$ and $b = -n+m-1$ with arbitrary limits:

$$\begin{aligned}
\sum_{k=0}^{n-m-1} \frac{[-n]_k}{[m-n-1]_k} &= \frac{1}{-m}\left(\frac{[-n]_{n-m}}{[m-n-1]_{n-m-1}} - \frac{1}{[m-n-1]_{-1}}\right) \\
&= \frac{1}{-m}\left(\frac{(-1)^{n-m}[n+n-m-1]_{n-m}}{(-1)^{n-m-1}[n-m+1+(n-m-2)]_{n-m-1}} - m + n\right) \\
&= \frac{1}{m}\left(\frac{[2n-m-1]_{n-m}}{[2n-2m-1]_{n-m-1}} - (n-m)\right) \\
&= \frac{1}{m}\left(\frac{[2n-m-1]_m}{[n-1]_{m-1}} - (n-m)\right).
\end{aligned}$$

Hence for the sum S, we get

$$\begin{aligned}
S &= \frac{1}{[m-1]_{m-1}} \cdot \frac{[n-1]_{m-1}}{1} \cdot \frac{1}{m}\left(\frac{[2n-m-1]_m}{[n-1]_{m-1}} - (n-m)\right) \\
&= \frac{1}{[m]_m}\left([2n-m-1]_m - [n-1]_m\right) \\
&= \frac{[2n-m-1]_m}{[m]_m} - \frac{[n-1]_m}{[m]_m} \\
&= \binom{2n-m-1}{2n-2m-1} - \binom{n-1}{m}.
\end{aligned}$$

8.7 The Number of Parentheses

Let P_n denote the number of ways we can place parentheses legally between n objects.[2] We want to establish the formula

$$P_n = \frac{1}{n}\binom{2n-2}{n-1}$$

from the obvious recursion

$$P_{n+1} = \sum_{k=1}^{n} P_k P_{n+1-k}.$$

[2]This example is from Graham, Knuth, and Patashnik, *Concrete Mathematics* [Graham et al. 94, p. 357].

Substitution shows that we need to prove the formula

$$\frac{1}{n+1}\binom{2n}{n} = \sum_{k=1}^{n} \frac{1}{k}\binom{2k-2}{k-1}\frac{1}{n+1-k}\binom{2n-2k}{n-k}.$$

Using the quotient we find that the sum is proportional to

$$\sum_{k=1}^{n} \binom{n+1}{k} \left[\tfrac{1}{2}\right]_k \left[\tfrac{1}{2}\right]_{n+1-k}.$$

Now, Chu–Vandermonde convolution says the sum from 0 to $n+1$ is 0, so we must have

$$\sum_{k=1}^{n} \binom{n+1}{k} \left[\tfrac{1}{2}\right]_k \left[\tfrac{1}{2}\right]_{n+1-k} = -2\left[\tfrac{1}{2}\right]_{n+1}. \tag{8.27}$$

Hence, we get

$$\begin{aligned}\sum_{k=1}^{n} \frac{1}{k}\binom{2k-2}{k-1}\frac{1}{n+1-k}\binom{2n-2k}{n-k} &= -\frac{[2n-2]_{n-1}}{(n+1)!\frac{1}{2}\left[\frac{1}{2}\right]_n} 2\left[\tfrac{1}{2}\right]_{n+1}\\ &= \frac{1}{n+1}\binom{2n}{n}.\end{aligned}$$

8.8 An Indefinite Sum of Type II(2,2,1)

For any complex numbers $a, b, c, d \in \mathbb{C}$ satisfying the condition that $a + b - c - d = 0$, we have the indefinite summation formula

$$\sum \frac{[a]_k[b]_k}{[c-1]_k[d-1]_k}\delta k = \frac{1}{(a-c)(b-c)} \cdot \frac{[a]_k[b]_k}{[c-1]_{k-1}[d-1]_{k-1}}.$$

We also have a generalization with *excess* p: For any complex numbers $a, b, c, d \in \mathbb{C}$ satisfying the condition that $p = a + b - c - d \in \mathbb{N}_0$, we have the indefinite summation formula

$$\sum \frac{[a]_k[b]_k}{[c-1]_k[d-1]_k}\delta k = \frac{[a]_k[b]_k}{[c-1]_{k-1}[d-1]_{k-1}} \sum_{j=0}^{p} \frac{[p]_j[k-c-1]_j}{[a-c]_{j+1}[b-c]_{j+1}}.$$

We proved these formulas as Corollary 6.2 and Theorem 6.1, formulas (6.15) and (6.13).

8.9 Transformations of Sums of Type II(2,2,z)

The standard form of a sum of type II$(2, 2, z)$ with natural limits is

$$T(a,b,n,z) = \sum_{k=0}^{n} \binom{n}{k} [a]_k [b]_{n-k} z^k, \tag{8.28}$$

with quotient

$$q_t(a,b,n,k) = \frac{(n-k)(a-k)}{(-1-k)(n-b-1-k)} z.$$

Explicit forms of the sum in (8.28) are in general not known. But for certain values of z, formulas exist with some restraints on the parameters (a, b). Furthermore, in general the possible values of z appear in groups of order 6. This is due to the following transformations.

Theorem 8.6 (Transformation Theorem). *The sums of the form* (8.28) *allow the following identities:*

$$T(a,b,n,z) = z^n T(b,a,n,1/z), \tag{8.29}$$
$$T(a,b,n,z) = (-1)^n T(a,n-a-b-1,n,1-z), \tag{8.30}$$
$$T(a,b,n,z) = (z-1)^n T(n-a-b-1,a,n,1/(1-z)), \tag{8.31}$$
$$T(a,b,n,z) = (1-z)^n T(n-a-b-1,b,n,z/(z-1)), \tag{8.32}$$
$$T(a,b,n,z) = (-z)^n T(b,n-a-b-1,n,1-1/z). \tag{8.33}$$

Proof: The formulas follow from repetition of two of them, (8.29) and (8.30). The former is obtained by reversing the direction of summation in (8.28). We only need to consider

$$\begin{aligned}
&(-1)^n T(a,n-a-b-1,n,1-z)\\
&=(-1)^n \sum_{k=0}^{n} \binom{n}{k} [a]_k [n-a-b-1]_{n-k} (1-z)^k\\
&=(-1)^n \sum_{k=0}^{n} \binom{n}{k} [a]_k [n-a-b-1]_{n-k} \sum_{j=0}^{k} \binom{k}{j} (-z)^j\\
&=(-1)^n \sum_{j=0}^{n} \binom{n}{j} (-z)^j \sum_{k=j}^{n} \binom{n-j}{k-j} [a]_k [n-a-b-1]_{n-k}\\
&=(-1)^n \sum_{j=0}^{n} \binom{n}{j} (-z)^j [a]_j \sum_{k=0}^{n-j} \binom{n-j}{k} [a-j]_k [n-a-b-1]_{n-j-k}
\end{aligned}$$

$$
\begin{aligned}
&=(-1)^n \sum_{j=0}^{n} \binom{n}{j} (-z)^j [a]_j [n-j-b-1]_{n-j} \\
&=(-1)^n \sum_{j=0}^{n} \binom{n}{j} (-z)^j [a]_j [-n+j+b+1+(n-j)-1]_{n-j} (-1)^{n-j} \\
&=\sum_{j=0}^{n} \binom{n}{j} z^j [a]_j [b]_{n-j} = T(a,b,n,z),
\end{aligned}
$$

where we have interchanged the order of summation and applied (2.9), (2.2), (2.1), and eventually the Chu–Vandermonde formula (8.2). □

This theorem tells us that with each possible value of z, there are up to five other formulas easily obtainable, giving the six values

$$z, \quad 1/z, \quad 1-z, \quad 1/(1-z), \quad z/(z-1), \quad 1-1/z.$$

The cases having fewer than six different possibilities are (a) $z=1$, giving at most $1-z=0$; (b) $z=1/z=-1$, related to $1-z=1-1/z=2$ and $1/(1-z)=z/(z-1)=\frac{1}{2}$, giving a group of 3; and (c) the sixth root of unity, $z=1/(1-z)=e^{i\frac{\pi}{3}}=1-1/z=\frac{1}{2}+i\frac{\sqrt{3}}{2}$, related to $1/z=1-z=z/(z-1)=e^{-i\frac{\pi}{3}}=\frac{1}{2}-i\frac{\sqrt{3}}{2}$, giving a group of 2 complex conjugates.

8.10 The Factors -1 and 2. Formulas of Kummer, Gauss, and Bailey

The formulas of types II$(2,2,-1)$ and II$(2,2,2)$ are due to E. E. Kummer in 1836 [Kummer 36], C. F. Gauss in 1813 [Gauss13], and W. N. Bailey in 1935 [Bailey 35]. As can be seen from the transformations (8.29)–(8.32), these formulas are closely related.

Consider a sum of the form

$$S(a,b,n,d) = \sum_{k=0}^{n} \binom{n}{k} [a,d]_k [b,d]_{n-k} (-1)^k,$$

with the quotient

$$q_k = \frac{(n-k)(\frac{a}{d}-k)}{(-1-k)(n-1-\frac{b}{d}-k)}(-1).$$

For $d = 0$ the sum reduces to the binomial theorem, and otherwise we can use (4.8) to reduce the it to the form of $d = 1$. The factor (-1) makes the sum alternating, so for $a = b$ and n odd we must have

$$S(a, a, 2n + 1, d) = -S(a, a, 2n + 1, d) = 0.$$

Theorem 8.7 (General Kummer Formula). *For* $a, b \in \mathbb{C}$,

$$S(a, b, n) = \sum_{j=0}^{\lfloor \frac{n}{2} \rfloor} \binom{b-a}{n-2j} [n]_{n-j} [a]_j (-1)^j.$$

Proof: Let us define

$$S = \sum_{k=0}^{n} \binom{n}{k} [a]_k [b]_{n-k} (-1)^k.$$

Then we apply (8.3) to the second factorial as

$$[b]_{n-k} = \sum_j \binom{n-k}{j-k} [a-k]_{j-k} [b-a+k]_{n-j},$$

and we obtain

$$S = \sum_k \sum_j \binom{n}{k} \binom{n-k}{j-k} (-1)^k [a]_k [a-k]_{j-k} [b-a+k]_{n-j}.$$

We apply (2.9) and (2.2) and interchange the order of summation to get

$$\begin{aligned} S &= \sum_j \sum_k \binom{n}{j} \binom{j}{k} (-1)^k [a]_j [b-a+k]_{n-j} \\ &= \sum_j \binom{n}{j} [a]_j \sum_k \binom{j}{k} [b-a+k]_{n-j} (-1)^k. \end{aligned}$$

Now (8.7) implies that this is zero for $0 \leq n - j < j$, but for $j \leq \frac{n}{2}$ we get

$$S = \sum_j \binom{n}{j} [a]_j (-1)^j [b-a]_{n-2j} [n-j]_j.$$

We apply (2.10) and (2.12) to write

$$\binom{n}{j} [n-j]_j = \binom{n-j}{j} [n]_j = \binom{n-j}{n-2j} [n]_j,$$

and then (2.10) again to write

$$\binom{n-j}{n-2j}[b-a]_{n-2j} = \binom{b-a}{n-2j}[n-j]_{n-2j},$$

and finally (2.2) to write

$$[n]_j[n-j]_{n-2j} = [n]_{n-j}.$$

The result is then

$$S = \sum_j \binom{b-a}{n-2j}[n]_{n-j}[a]_j(-1)^j.$$

□

In general this theorem is the only formula known, but for two special cases there are improvements. If either $b-a$ or $b+a$ are integers, we can shorten the sum.

In the first case we have the quasi-symmetric Kummer formula with $p = b - a \in \mathbb{Z}$.

Theorem 8.8 (Quasi-Symmetric Kummer Formula). *For $a \in \mathbb{C}$ and $p \in \mathbb{Z}$,*

$$S(a, a+p, n) = \sigma(p)^n \sum_{j=\lceil \frac{n-|p|}{2} \rceil}^{\lfloor \frac{n}{2} \rfloor} \binom{|p|}{n-2j}[n]_{n-j}[a+(p\wedge 0)]_j(-1)^j. \quad (8.34)$$

Proof: Let $p \geq 0$. Then the change to the natural limits gives the formula (8.34). The general formula is obtained by reversing the order of summation. □

The special case of difference zero is the symmetric Kummer identity.

Corollary 8.9 (Symmetric Kummer Identity). *For $a \in \mathbb{C}$,*

$$\sum_{k=0}^{n} \binom{n}{k}[a]_k[a]_{n-k}(-1)^k = \begin{cases} [n]_m[a]_m(-1)^m & \text{for } n = 2m, \\ 0 & \text{for } n \text{ odd.} \end{cases}$$

By division with $[n]_n = [n]_m[m]_m$ we can rewrite the formula (8.34) to look analogous to (8.1):

Corollary 8.10 (Quasi-Symmetric Kummer Formula).

$$\sum_{k=0}^{n} \binom{x}{k}\binom{x+p}{n-k}(-1)^k = \sigma(p)^n \sum_{j=\lceil \frac{n-|p|}{2} \rceil}^{\lfloor \frac{n}{2} \rfloor} \binom{|p|}{n-2j}\binom{x+(p\wedge 0)}{j}(-1)^j. \quad (8.35)$$

In the second case, $a+b \in \mathbb{Z}$, we have the different quasi-balanced and balanced Kummer formulas. To prove these we need the following:

Theorem 8.11 (First Quasi-Balanced Kummer Formula).
For $a \in \mathbb{C}$ *and* $0 \leq n \geq p \in \mathbb{Z}$,

$$\sum_{k=0}^{n}\binom{n}{k}[n-p+a]_k[n-a]_{n-k}(-1)^k$$
$$= \frac{2^{n-(p\vee 0)}}{[n-p]_{-(p\wedge 0)}}\sum_{j=0}^{|p|}\binom{|p|}{j}(-\sigma(p))^{p+j}$$
$$\times [n-a]_j[n-p+a]_{|p|-j}[n-j-1-a,2]_{n-(p\vee 0)}. \qquad (8.36)$$

Proof: We apply (2.1) to the first factorial to obtain

$$S = \sum_{k=0}^{n}\binom{n}{k}[n-p+a]_k[n-a]_{n-k}(-1)^k$$
$$= \sum_{k}\binom{n}{k}[n-a]_{n-k}[p-n-a+k-1]_k.$$

The product runs from $n-a$ to $p-n-a$ with the exception of the factors in the factorial $[k-a]_{n-p+1}$. Therefore, using (2.2) twice, we can write,

$$S = \sum_{k}\binom{n}{k}\frac{[n-a]_{n-k}[k-a]_{n-p+1}[p-n-a+k-1]_k}{[k-a]_{n-p+1}}$$
$$= \sum_{k}\binom{n}{k}\frac{[n-a]_{2n-p+1}}{[k-a+p-n+(n-p)]_{n-p+1}}.$$

We now apply (8.11) to the denominator with $c = k-a+p-n$ and $n = n-p$ to get

$$S = [n-a]_{2n-p+1}\sum_{k}\binom{n}{k}\frac{1}{[n-p]_{n-p}}\sum_{j}\binom{n-p}{j}\frac{(-1)^j}{k-a+p-n+j}.$$

We substitute $i = k+j$ as summation variable in the second sum and get

$$S = \frac{[n-a]_{2n-p+1}}{[n-p]_{n-p}}\sum_{k}\binom{n}{k}\sum_{i}\binom{n-p}{i-k}\frac{(-1)^{i-k}}{p-n-a+i}.$$

Then we interchange the order of summation and receive

$$S = \frac{[n-a]_{2n-p+1}}{[n-p]_{n-p}}\sum_{i}\frac{(-1)^i}{p-n-a+i}\sum_{k}\binom{n}{k}\binom{n-p}{i-k}(-1)^k.$$

Now the integral quasi-symmetric Kummer formula (8.35) applies to the second sum (with the sign of p changed). Hence, we get

$$S = \frac{[n-a]_{2n-p+1}}{[n-p]_{n-p}} \sum_i \frac{(-1)^i}{p-n-a+i} (-\sigma(p))^i \sum_j \binom{|p|}{i-2j} \times \binom{n-(p\vee 0)}{j} (-1)^j.$$

Next we interchange the order of summation and get

$$S = \frac{[n-a]_{2n-p+1}}{[n-p]_{n-p}} \sum_j \binom{n-(p\vee 0)}{j} (-1)^j \sum_i \binom{|p|}{i-2j} \frac{(\sigma(p))^i}{p-n-a+i}.$$

Now substitute $k = i - 2j$ as the summation variable in the second sum and get

$$S = \frac{[n-a]_{2n-p+1}}{[n-p]_{n-p}} \sum_j \binom{n-(p\vee 0)}{j} (-1)^j \sum_k \binom{|p|}{k} \frac{(\sigma(p))^k}{p-n-a+k+2j}.$$

Again we interchange the order of summation while dividing the denominator by 2

$$S = \frac{[n-a]_{2n-p+1}}{[n-p]_{n-p}} \sum_k \binom{|p|}{k} \frac{(\sigma(p))^k}{2} \sum_j \binom{n-(p\vee 0)}{j} \frac{(-1)^j}{\frac{p-n-a+k}{2}+j}.$$

Now we can again apply (8.11) with $c = \frac{p-n-a+k}{2}$, $n = n-(p\vee 0)$ to the inner sum. It becomes

$$S = \frac{[n-a]_{2n-p+1}}{[n-p]_{n-p}} \sum_k \binom{|p|}{k} \frac{(\sigma(p))^k}{2} \cdot \frac{[n-(p\vee 0)]_{n-(p\vee 0)}}{\left[\frac{p-n-a+k}{2}+n-(p\vee 0)\right]_{n-(p\vee 0)+1}}.$$

The next step is to double each factor in the factorial of the denominator, which also doubles the step size, then cancel common factors of the factorials before the sum, and finally we get

$$S = \frac{2^{n-(p\vee 0)}}{[n-p]_{-(p\wedge 0)}} [n-a]_{2n-p+1} \sum_k \binom{|p|}{k} \frac{(\sigma(p))^k}{[n-a+k-|p|,2]_{n-(p\vee 0)+1}}.$$

We reverse the direction of summation and get

$$S = \frac{2^{n-(p\vee 0)}}{[n-p]_{-(p\wedge 0)}} \sum_k \binom{|p|}{k} (\sigma(p))^{p-k} \frac{[n-a]_{2n-p+1}}{[n-a-k,2]_{n-(p\vee 0)+1}}.$$

We use (2.2) to split the numerator as

$$\begin{aligned}&[n-a]_{2n-p+1}\\&\quad=[n-a]_k[n-a-k]_{2n-2(p\vee 0)+1}[-a-n-k+2(p\vee 0)-1]_{2(p\vee 0)-p-k},\end{aligned}$$

and (2.7) to split

$$[n-a-k]_{2n-2(p\vee 0)+1}=[n-a-k,2]_{n-(p\vee 0)+1}[n-a-k-1,2]_{n-(p\vee 0)},$$

and finally (2.1) to write the last factorial as

$$[-a-n-k+2(p\vee 0)-1]_{2(p\vee 0)-p-k}=(-1)^{p-k}[a+n-p]_{|p|-k}.$$

Then the result follows. □

Theorem 8.12 (Balanced Kummer Identity). *For $a\in\mathbb{C}$,*

$$\sum_{k=0}^{n}\binom{n}{k}[n+a]_k[n-a]_{n-k}(-1)^k=2^n[n-1-a,2]_n. \tag{8.37}$$

Proof: We just have to apply (8.36) for $p=0$. □

Theorem 8.13 (Second Quasi-Balanced Kummer Formula). *For $a\in\mathbb{C}$ and $p\in\mathbb{Z}$,*

$$\begin{aligned}&\sum_{k=0}^{n}\binom{n}{k}[n-p+a]_k[n-a]_{n-k}(-1)^k\\&\quad=\frac{2^{n-(p\vee 0)}}{[n-p]_{-(p\wedge 0)}}\sum_{j=0}^{|p|}\binom{|p|}{j}\sigma(p)^{p+j}[-a+n+j-1,2]_{n-(p\wedge 0)}.\end{aligned} \tag{8.38}$$

Proof: We define

$$S(n,a,p):=\sum_{k=0}^{n}\binom{n}{k}[n-p+a]_k[n-a]_{n-k}(-1)^k.$$

We apply (2.4) to the second factorial to split the sum into two, and then formula (2.7) to the second sum to get

$$\begin{aligned}S(n,a,p)&=\sum_{k=0}^{n}\binom{n}{k}[(n-(p+1))+(a+1)]_k[n-1-a]_{n-k}(-1)^k\\&\quad+n\sum_{k=0}^{n-1}\binom{n-1}{k}[((n-1)-(p-1))+a]_k[n-1-a]_{n-1-k}(-1)^k\\&=S(n,a+1,p+1)+nS(n-1,a,p-1).\end{aligned}$$

Next we apply (2.4) to the first factorial to split the sum in two, and then the formula (2.7) to the second sum to get

$$\begin{aligned} S(n,a,p) &= \sum_{k=0}^{n} \binom{n}{k} [(n-(p+1))+a]_k [n-a]_{n-k} (-1)^k \\ &\qquad + n \sum_{k=1}^{n} \binom{n-1}{k-1} [((n-1)-(p-1))+(a-1)]_{k-1} \\ &\qquad\qquad \times [n-1-(a-1)]_{n-1-(k-1)} (-1)^k \\ &= S(n,a,p+1) - nS(n-1,a-1,p-1). \end{aligned}$$

By eliminating respectively the second and the first terms of the right sides of the two formulas, we obtain two useful recursions:

$$\begin{aligned} 2S(n,a,p) &= S(n,a-1,p-1) + S(n,a,p-1), \\ 2(n+1)S(n,a,p) &= S(n+1,a,p+1) - S(n+1,a+1,p+1). \end{aligned}$$

Repeating the first formula p times or the second $-p$ times, then using the formula (8.37), valid for $p=0$, and cancelling common powers of 2, yields the form (8.38), after reversing the direction of summation. □

Corollary 8.14. *For arbitrary c, arbitrary $n, p \in \mathbb{N}_0$ and $\delta = \pm 1$ we have*

$$\begin{aligned} &\sum_{j=0}^{p} \binom{p}{j} (\delta)^{p+j} [c+j,2]_{n+p} \\ &\qquad = \sum_{j=0}^{p} \binom{p}{j} (-\delta)^{p+j} [c-1]_j [2n+p-1-c]_{p-j} [c-j,2]_n. \end{aligned}$$

Proof: Compare the right sides of (8.36) and (8.38) and cancel the common factors. □

Related to the factor $z = -1$ is the factor 2, which appears in well-known formulas, e.g., the formulas of Gauss from 1813 [Gauss13], and Bailey from 1935 [Bailey 35], which are the following formulas for the choice $p = 0$.

Theorem 8.15 (Generalized Gauss Identity). *For $a \in \mathbb{C}$, $p \in \mathbb{Z}$, and $n \in \mathbb{N}_0$,*

$$\sum_{k=0}^{n} \binom{n}{k} [a]_k [n-p-1-2a]_{n-k} 2^k$$
$$= (-\sigma(p))^n \sum_{j=\lceil \frac{n-|p|}{2} \rceil}^{\lfloor \frac{n}{2} \rfloor} \binom{|p|}{n-2j} [n]_{n-j} [a+(p \wedge 0)]_j (-1)^j. \tag{8.39}$$

The Gauss quotient is

$$q_g = \frac{(n-k)(a-k)}{(-1-k)(2a+p-k)} 2,$$

making it type II(2, 2, 2).

Theorem 8.16 (Generalized Bailey Identity). *For $a \in \mathbb{C}$, $p \in \mathbb{Z}$, and $n \in \mathbb{N}_0$,*

$$2^{(p-n)\vee 0} [n-p]_{-(p\wedge 0)} \sum_{k=0}^{n} \binom{n}{k} [a]_k [(p-n-1)]_{n-k} 2^k$$
$$= (-1)^n 2^{(n-(p\vee 0))\vee 0} \sum_{j=0}^{|p|} \binom{|p|}{j} \sigma(p)^{p+j} [-a+2n-p+j-1, 2]_{n-(p\wedge 0)}. \tag{8.40}$$

The Bailey quotient is

$$q_b = \frac{(n-k)(a-k)}{(-1-k)(2n-p-k)} 2,$$

making it type II(2, 2, 2).

Proof: We apply (8.30) to (8.34) to obtain the generalized Gauss formula (8.39).

We replace a by $a+p-n$ in (8.38) to obtain as the left side

$$\sum_{k=0}^{n} \binom{n}{k} [a]_k [2n+p-a]_{n-k} (-1)^k. \tag{8.41}$$

When we apply (8.30) to (8.41), we obtain the generalized Bailey formula (8.40). □

8.11 The Factor $\frac{1}{2} + i\frac{\sqrt{3}}{2}$

Let the three parameters in the transformation formulas (8.29)–(8.32) be equal: $a = b = n - a - b - 1$, i.e., $a = b = \frac{n-1}{3}$. Then the transformation (8.31) yields, with $\rho = \frac{1}{2} + i\frac{\sqrt{3}}{2}$,

$$T(\tfrac{n-1}{3}, \tfrac{n-1}{3}, n, \rho) = (\rho - 1)^n T(\tfrac{n-1}{3}, \tfrac{n-1}{3}, n, \rho) = \rho^{2n} T(\tfrac{n-1}{3}, \tfrac{n-1}{3}, n, \rho).$$

This means that we get 0, except when $n = 3m$, because $\rho^6 = 1$. Furthermore, for m even the sum must be real, while it is purely imaginary for m odd.

In fact, we have the formula

$$\sum_{k=0}^{n} \binom{n}{k} \left[\tfrac{n-1}{3}\right]_k \left[\tfrac{n-1}{3}\right]_{n-k} \rho^k = \begin{cases} 0 & 3 \nmid n, \\ \dfrac{[3m]_{2m} \left[-\frac{2}{3}\right]_m i^m}{(\sqrt{27})^m} & n = 3m. \end{cases} \qquad (8.42)$$

This is a special case of number 15.1.31 in [Abramowitz and Stegun 65], which has the following three forms as special cases:

$$\sum_{k=0}^{n} \binom{n}{k} [3n+1]_k [-n-1]_{n-k} \rho^k = \left[-\tfrac{2}{3}\right]_n \left(-i\sqrt{27}\right)^n,$$

$$\sum_{k=0}^{n} \binom{n}{k} [-n-1]_k [3n+1]_{n-k} \rho^k = \left[-\tfrac{2}{3}\right]_n \left(-i\sqrt{27}\rho\right)^n,$$

$$\sum_{k=0}^{n} \binom{n}{k} [-n-1]_k [-n-1]_{n-k} \rho^k = \left[-\tfrac{2}{3}\right]_n \left(i\sqrt{27}\bar{\rho}\right)^n.$$

8.12 Sums of Type II(2,2,z)

From formulas (12.21)–(12.24), we know the sum for the four values of the constants in the factorial, $n \pm \frac{1}{2}$, and any choice of z.

$$\sum_{k=0}^{n} \binom{n}{k} \left[n - \tfrac{1}{2}\right]_k \left[n - \tfrac{1}{2}\right]_{n-k} z^k$$
$$= [2n-1]_n \left(\left(\tfrac{\sqrt{z}+1}{2}\right)^{2n} + \left(\tfrac{\sqrt{z}-1}{2}\right)^{2n} \right), \qquad (8.43)$$

$$\sum_{k=0}^{n} \binom{n}{k} \left[n+\tfrac{1}{2}\right]_k \left[n-\tfrac{1}{2}\right]_{n-k} z^k$$
$$= [2n]_n \left(\left(\tfrac{\sqrt{z}+1}{2}\right)^{2n+1} - \left(\tfrac{\sqrt{z}-1}{2}\right)^{2n+1}\right), \tag{8.44}$$
$$\sum_{k=0}^{n} \binom{n}{k} \left[n+\tfrac{1}{2}\right]_k \left[n+\tfrac{1}{2}\right]_{n-k} z^k$$
$$= \frac{[2n+1]_n}{\sqrt{z}} \left(\left(\tfrac{\sqrt{z}+1}{2}\right)^{2n+2} - \left(\tfrac{\sqrt{z}-1}{2}\right)^{2n+2}\right). \tag{8.45}$$

The formulas (8.43)–(8.45) are in Gould's table [Gould 72b, 1.38, 1.39 and 1.70, 1.71]. Some of the formulas above for the choices $z = 5, 9, 25$ are repeated in Gould's table as numbers 1.74, 1.77 and 1.69, respectively. To recognize them it may be necessary to apply some of the transformations (8.13) and (8.14). For completeness we will give the formulas obtained by these transformations in the nine other possible forms. The first one (8.46) appears as number 1.64 in Gould's table.

$$\sum_{k=0}^{n} \binom{n}{k} \left[n-\tfrac{1}{2}\right]_k [-n]_{n-k} z^k$$
$$= [-n]_n \left(\left(\tfrac{\sqrt{1-z}+1}{2}\right)^{2n} + \left(\tfrac{\sqrt{1-z}-1}{2}\right)^{2n}\right), \tag{8.46}$$
$$\sum_{k=0}^{n} \binom{n}{k} [-n]_k \left[n-\tfrac{1}{2}\right]_{n-k} z^k$$
$$= [-n]_n \left(\left(\tfrac{\sqrt{z}+\sqrt{z-1}}{2}\right)^{2n} + \left(\tfrac{\sqrt{z}-\sqrt{z-1}}{2}\right)^{2n}\right), \tag{8.47}$$
$$\sum_{k=0}^{n} \binom{n}{k} \left[n+\tfrac{1}{2}\right]_k [-n-1]_{n-k} z^k$$
$$= [-n-1]_n \left(\left(\tfrac{\sqrt{1-z}+1}{2}\right)^{2n+1} - \left(\tfrac{\sqrt{1-z}-1}{2}\right)^{2n+1}\right), \tag{8.48}$$
$$\sum_{k=0}^{n} \binom{n}{k} [-n-1]_k \left[n+\tfrac{1}{2}\right]_{n-k} z^k$$
$$= \tfrac{[-n-1]_n}{\sqrt{z}} \left(\left(\tfrac{\sqrt{z-1}+\sqrt{z}}{2}\right)^{2n+1} - \left(\tfrac{\sqrt{z-1}-\sqrt{z}}{2}\right)^{2n+1}\right), \tag{8.49}$$

$$\sum_{k=0}^{n} \binom{n}{k} \left[n-\tfrac{1}{2}\right]_k [-n-1]_{n-k}\, z^k = \tfrac{[-n-1]_n}{\sqrt{1-z}} \left(\left(\tfrac{1+\sqrt{1-z}}{2}\right)^{2n+1} - \left(\tfrac{1-\sqrt{1-z}}{2}\right)^{2n+1} \right), \tag{8.50}$$

$$\sum_{k=0}^{n} \binom{n}{k} [-n-1]_k \left[n-\tfrac{1}{2}\right]_{n-k} z^k = \tfrac{[-n-1]_n}{\sqrt{z-1}} \left(\left(\tfrac{\sqrt{z}+\sqrt{z-1}}{2}\right)^{2n+1} - \left(\tfrac{\sqrt{z}-\sqrt{z-1}}{2}\right)^{2n+1} \right), \tag{8.51}$$

$$\sum_{k=0}^{n} \binom{n}{k} \left[n-\tfrac{1}{2}\right]_k \left[n+\tfrac{1}{2}\right]_{n-k} z^k = \tfrac{[2n]_n}{\sqrt{z}} \left(\left(\tfrac{1+\sqrt{z}}{2}\right)^{2n+1} - \left(\tfrac{1-\sqrt{z}}{2}\right)^{2n+1} \right), \tag{8.52}$$

$$\sum_{k=0}^{n} \binom{n}{k} \left[n+\tfrac{1}{2}\right]_k [-n-2]_{n-k}\, z^k = \tfrac{[-n-2]_n}{\sqrt{1-z}} \left(\left(\tfrac{\sqrt{1-z}+1}{2}\right)^{2n+2} - \left(\tfrac{\sqrt{1-z}-1}{2}\right)^{2n+2} \right), \tag{8.53}$$

$$\sum_{k=0}^{n} \binom{n}{k} [-n-2]_k \left[n+\tfrac{1}{2}\right]_{n-k} z^k = \tfrac{[-n-2]_n}{\sqrt{z}\sqrt{z-1}} \left(\left(\tfrac{\sqrt{z-1}+\sqrt{z}}{2}\right)^{2n+2} - \left(\tfrac{\sqrt{z-1}-\sqrt{z}}{2}\right)^{2n+2} \right). \tag{8.54}$$

Remark 8.17. In the formulas the choice of the square roots shall be consistent throughout the right side.

It is possible to generalize these formulas by introducing a parameter, $m \in \mathbb{Z}$, to add to one of or both of the arguments. For $m < 0$ we only know some recurrence formulas, but for $m \in \mathbb{N}_0$ we can give exact expressions. For example, replacing z by various expressions in r, we have

$$\begin{aligned}\sum_{k=0}^{n} \binom{n}{k} \left[n-\tfrac{1}{2}\right]_k \left[n+m+\tfrac{1}{2}\right]_{n-k} \left(\frac{1+r}{1-r}\right)^{2k} &= \frac{[2n]_{n-m}}{(r-1)^{2n}} \sum_{k=0}^{m} \frac{[2n+2m]_k[-m]_{m-k}}{(1+r)^{2m+1-k}} \\ &\times \left(\left(\binom{m}{k} r - \binom{m-1}{k} \right) r^{2n+2m-k} + \binom{m}{k} - r\binom{m-1}{k} \right),\end{aligned} \tag{8.55}$$

$$\sum_{k=0}^{n}\binom{n}{k}\left[n+\tfrac{1}{2}\right]_k\left[n+m+\tfrac{1}{2}\right]_{n-k}\left(\frac{1+r}{1-r}\right)^{2k}$$
$$=\frac{[2n+1]_{n-m}}{(r-1)^{2n+1}}\sum_{k=0}^{m}\frac{[2n+2m+1]_k[-m]_{m-k}}{(1+r)^{2m+1-k}}$$
$$\times\left(\left(\binom{m}{k}r-\binom{m-1}{k}\right)r^{2n+2m+1-k}+\binom{m}{k}-r\binom{m-1}{k}\right), \quad (8.56)$$

$$\sum_{k=0}^{n}\binom{n}{k}\left[n-\tfrac{1}{2}\right]_k\left[-n-m-1\right]_{n-k}\left(\frac{-4r}{(1-r)^2}\right)^{k}$$
$$=\frac{[2n]_{n-m}(-1)^n}{(r-1)^{2n}}\sum_{k=0}^{m}\frac{[2n+2m]_k[-m]_{m-k}}{(1+r)^{2m+1-k}}$$
$$\times\left(\left(\binom{m}{k}r-\binom{m-1}{k}\right)r^{2n+2m-k}+\binom{m}{k}-r\binom{m-1}{k}\right), \quad (8.57)$$

$$\sum_{k=0}^{n}\binom{n}{k}\left[n+\tfrac{1}{2}\right]_k\left[-n-m-2\right]_{n-k}\left(\frac{-4r}{(1-r)^2}\right)^{k}$$
$$=\frac{[2n+1]_{n-m}(-1)^n}{(r-1)^{2n+1}}\sum_{k=0}^{m}\frac{[2n+2m+1]_k[-m]_{m-k}}{(1+r)^{2m+1-k}}$$
$$\times\left(\left(\binom{m}{k}r-\binom{m-1}{k}\right)r^{2n+2m+1-k}+\binom{m}{k}-r\binom{m-1}{k}\right), \quad (8.58)$$

$$\sum_{k=0}^{n}\binom{n}{k}\left[n+m+\tfrac{1}{2}\right]_k\left[-n-m-1\right]_{n-k}\left(\frac{-4r}{(1+r)^2}\right)^{k}$$
$$=\frac{[2n]_{n-m}(-1)^n}{(r+1)^{2n}}\sum_{k=0}^{m}\frac{[2n+2m]_k[-m]_{m-k}}{(1+r)^{2m+1-k}}$$
$$\times\left(\left(\binom{m}{k}r-\binom{m-1}{k}\right)r^{2n+2m-k}+\binom{m}{k}-r\binom{m-1}{k}\right), \quad (8.59)$$

$$\sum_{k=0}^{n}\binom{n}{k}\left[n+m+\tfrac{1}{2}\right]_k\left[-n-m-2\right]_{n-k}\left(\frac{-4r}{(1+r)^2}\right)^{k}$$
$$=\frac{[2n+1]_{n-m}(-1)^n}{(r+1)^{2n}(r-1)}\sum_{k=0}^{m}\frac{[2n+2m+1]_k[-m]_{m-k}}{(1+r)^{2m+1-k}}$$
$$\times\left(\left(\binom{m}{k}r-\binom{m-1}{k}\right)r^{2n+2m+1-k}+\binom{m}{k}-r\binom{m-1}{k}\right). \quad (8.60)$$

We prove these formulas by using the transformations (8.13)–(8.14) and, in the forms of (8.57)–(8.58), the transformations (5.20) and (5.21). These formulas are then joined by the formula (5.19), so all that remains to be proved is the following:

Theorem 8.18. *For $r = 1$ and any $m \in \mathbb{N}_0$ we have the formula*

$$\sum_{j=0}^{\lfloor \frac{n}{2} \rfloor} [j]_m \binom{n-j}{j} (-1)^j 2^{n-2j} = (-1)^m m! \binom{n+1}{2m+1}. \tag{8.61}$$

For any $r \in \mathbb{C} \setminus \{1\}$ and any $m \in \mathbb{N}_0$ we have the formula

$$\sum_{j=0}^{\lfloor \frac{n}{2} \rfloor} [j]_m \binom{n-j}{j} (-r)^j (1+r)^{n-2j} = (-r)^m \sum_{k=0}^{m} \frac{[n]_k [-m]_{m-k}}{(1-r)^{2m+1-k}}$$
$$\times \left(\left(\binom{m}{k} + r \binom{m-1}{k} \right) - \left(\binom{m}{k} r + \binom{m-1}{k} \right) (-1)^k r^{n-k} \right). \tag{8.62}$$

To prove this formula we will find useful a more general lemma, valid for $m \in \mathbb{Z}$, establishing a recursion for the kind of sums we consider.

Lemma 8.19. *The function defined for $n \in \mathbb{N}_0$, $m \in \mathbb{Z}$, and $r \in \mathbb{C}$ by*

$$f(n,m,r) = \sum_{j=0}^{\lfloor \frac{n}{2} \rfloor} [j]_m \binom{n-j}{j} (-r)^j (1+r)^{n-2j} \tag{8.63}$$

satisfies the following recursion in n and m:

$$f(n+2,m,r) - (1+r) f(n+1,m,r) + r f(n,m,r) = -mr f(n,m-1,r). \tag{8.64}$$

Proof: We split the binomial coefficient with (2.7) into two to write

$$\begin{aligned} f(n+2,m,r) &= \sum_{j=0}^{\lfloor \frac{n}{2} \rfloor + 1} [j]_m \binom{n+1-j}{j} (-r)^j (1+r)^{n+2-2j} \\ &\quad + \sum_{j=1}^{\lfloor \frac{n}{2} \rfloor + 1} [j]_m \binom{n+1-j}{j-1} (-r)^j (1+r)^{n+2-2j} \\ &= (1+r) f(n+1,m,r) \\ &\quad + \sum_{j=0}^{\lfloor \frac{n}{2} \rfloor} [j+1]_m \binom{n-j}{j} (-r)^{j+1} (1+r)^{n-2j}. \end{aligned}$$

Now apply (2.4) to split the factorial to establish formula (8.64). □

Proof of Theorem 8.18: The difference equation (8.64) has the characteristic roots r and 1. Hence, by Theorem 4.1, the homogeneous equation is solved by the functions

$$\alpha r^n + \beta \text{ for } r \neq 1,$$
$$\alpha n + \beta \text{ for } r = 1.$$

In particular, it follows from the equations $f(0,0,r) = 1$ and $f(1,0,r) = 1 + r$ that the solutions for $m = 0$ are

$$f(n,0,r) = \frac{r^{n+1} - 1}{r - 1} \text{ for } r \neq 1,$$
$$f(n,0,1) = n + 1.$$

The solutions can now be found by splitting the operator and applying (4.10) repeatedly. This procedure must give the solutions in the form

$$f(n,m,r) = p_m(n,r)r^n + q_m(n,r) \text{ for } r \neq 1, \tag{8.65}$$
$$f(n,m,1) = s_m(n), \tag{8.66}$$

where p_m, q_m, and s_m are polynomials in n of degrees m and $2m+1$ respectively.

Let us consider the special case $r = 1$. Then we can write

$$s_m(n) = \sum_{k=0}^{2m+1} \gamma_k^m [n]_k,$$

for which the lemma tells us that

$$\sum_{k=2}^{2m+1} \gamma_k^m [k]_2 [n]_{k-2} = -m \sum_{k=0}^{2m-1} \gamma_k^{m-1} [n]_k.$$

This gives the recursion in the coefficients

$$\gamma_k^m = -m \frac{\gamma_{k-2}^{m-1}}{[k]_2} \text{ for } k > 1.$$

Hence, we get $\gamma_2^1 = -\frac{1}{2}$ and $\gamma_3^1 = -\frac{1}{6}$, so that we find

$$f(n,1,1) = -\tfrac{1}{6}[n]_3 - \tfrac{1}{2}[n]_2 + \gamma_1^1 n + \gamma_0^1.$$

The last two terms must vanish, so the final formula is

$$f(n,1,1) = -\binom{n+1}{3}.$$

Induction now proves (8.61), but this case also follows from Chu–Vandermonde, (8.2).

Substitution of (8.65) into (8.64) yields, after division by the appropriate power of $r \neq 0$,

$$rp_m(n+2,r) - (1+r)p_m(n+1,r) + p_m(n,r) = -mp_{m-1}(n,r), \quad (8.67)$$
$$\tfrac{1}{r}q_m(n+2,r) - (1+\tfrac{1}{r})q_m(n+1,r) + q_m(n,r) = -mq_{m-1}(n,r). \quad (8.68)$$

We can write the two polynomials as

$$p_m(n,r) = \sum_{k=0}^{m} \alpha_k^m(r)[n]_k, \quad (8.69)$$
$$q_m(n,r) = \sum_{k=0}^{m} \beta_k^m(r)[n]_k, \quad (8.70)$$

and note that we already know

$$p_0(n,r) = \frac{r}{r-1}, \quad q_0(n,r) = \frac{1}{r-1} = -p_0\left(n, \tfrac{1}{r}\right).$$

Now substitution of (8.69) into (8.67) gives

$$\alpha_k^m(r) = \frac{1}{1-r}\left(\frac{m}{k}\alpha_{k-1}^{m-1}(r) + r(k+1)\alpha_{k+1}^m(r)\right). \quad (8.71)$$

For $m > 0$ we have the equations

$$\alpha_0^m(r) + \beta_0^m(r) = 0, \quad (8.72)$$
$$\left(\alpha_1^m(r) + \alpha_0^m(r)\right) r + \beta_1^m(r) + \beta_0^m(r) = 0. \quad (8.73)$$

If we assume that

$$\beta_1^m(r) = -\alpha_1^m\left(\tfrac{1}{r}\right),$$

then the solution of the simultaneous equations (8.72) and (8.73) gives

$$\beta_0^m(r) = -\alpha_0^m\left(\tfrac{1}{r}\right).$$

The similarity of equations (8.67) and (8.68) insures that we have the general identities

$$\beta_k^m(r) = -\alpha_k^m\left(\tfrac{1}{r}\right),$$
$$q_m(n,r) = -p_m\left(n, \tfrac{1}{r}\right).$$

The recursion (8.71) shows that

$$\alpha_m^m(r) = \frac{1}{1-r}\alpha_{m-1}^{m-1}(r) = \frac{r}{(1-r)^{m+1}}.$$

Now we define

$$\sigma_k^m(r) = \frac{(1-r)^{2m+1-k}}{[m]_{m-k}}\alpha_k^m(r).$$

Substitution into (8.71) gives the recursion for this new function,

$$\sigma_k^m(r) = \sigma_{k-1}^{m-1}(r) + r\sigma_{k+1}^m(r). \tag{8.74}$$

In particular, we have $\sigma_m^m(r) = r$. Induction proves that $\sigma_k^m(r)$ is a polynomial in r with only two terms, namely

$$\sigma_k^m(r) = \phi_k^m r^{m-k+1} + \psi_k^m r^{m-k}. \tag{8.75}$$

Substitution of (8.75) into (8.74) yields the very same recurrence for the coefficients, namely

$$\phi_k^m = \phi_{k-1}^{m-1} + \phi_{k+1}^m,$$

which has as solution some kind of binomial coefficients. We find the solutions to be

$$\phi_k^m = \binom{2m-k-1}{m},$$

$$\psi_k^m = \binom{2m-k-1}{m-1}.$$

Substitution backwards eventually yields

$$p_m(n,r) = -\sum_{k=0}^{m}\left(\binom{m}{k}r + \binom{m-1}{k}\right)\frac{[-m]_{m-k}(-r)^{m-k}}{(1-r)^{2m+1-k}}[n]_k, \tag{8.76}$$

$$q_m(n,r) = \sum_{k=0}^{m}\left(\binom{m}{k} + r\binom{m-1}{k}\right)\frac{[-m]_{m-k}(-r)^{m}}{(1-r)^{2m+1-k}}[n]_k. \tag{8.77}$$

Substitution of the formulas (8.76) and (8.77) into (8.65) gives the formula in the theorem, (8.62). □

Remark 8.20. The theorem in particular gives the formula for $m = 0$, and gives an easy way to find the formula for $m = -1$:

$$f(n,-1,r) = \frac{1}{(n+2)r}\left((1+r)^n - r^n - 1\right). \tag{8.78}$$

The recursion (8.63) in the lemma allows the computation of formulas for $m < -1$ as soon as the formula for $m = -1$ is known.

Proof of (8.78): Consider the formula for $m = 0$:

$$f(n,0,r) = \sum_{j=0}^{\lfloor \frac{n}{2} \rfloor} \binom{n-j}{j} (-r)^j (1+r)^{n-2j} = \frac{r^{n+1}-1}{r-1}.$$

Take the first term of the sum and remove the two highest factors of the binomial coefficient to get

$$f(n,0,r) = (1+r)^n + \sum_{j=1}^{\lfloor \frac{n}{2} \rfloor} \frac{n-j}{j} \binom{n-j-1}{j-1} (-r)^j (1+r)^{n-2j}.$$

In the sum, change the variable to $j-1$ to get

$$f(n,0,r) = (1+r)^n - r \sum_{j=0}^{\lfloor \frac{n}{2} \rfloor - 1} \left(\frac{n}{j+1} - 1 \right) \binom{n-2-j}{j} (-r)^j (1+r)^{n-2-2j},$$

which can be written as

$$f(n,0,r) = (1+r)^n - rf(n-2,-1,r) + rf(n-2,0,r).$$

This proves the formula (8.78), which remains valid for $r = 1$. □

8.13 The General Difference Equation

The only general result about sums of the form (8.28) is that the function $T(a,b,n,z)$ must satisfy the following second order difference equation in the variable n:

$$\left(\mathbf{E}^2 + ((1+z)(n+1) - za - b)\,\mathbf{E} + z(n-a-b)(n+1)\mathbf{I}\right) T(a,b,n,z) = 0. \tag{8.79}$$

For fixed a, b, z the equation has coefficients which are polynomials in n of degree 2, so no general solution is obtainable. For special choices of a, b, and z it is, of course, possible to solve the equation, but for all such choices which we have been able to figure out, we get some of the formulas mentioned above. This formula is derived by the Zeilberger algorithm, cf. (13.7).

Chapter 9

Sums of Type II(3,3,z)

9.1 The Pfaff–Saalschütz and Dixon Formulas

The most famous identities of type II$(3, 3, z)$ are the Pfaff–Saallschütz identity, first discovered by J. F. Pfaff in 1797 [Pfaff 97] and later reformulated by L. Saalschütz in 1890 [Saalschütz 90], and the two identities due to A. C. Dixon in 1903 [Dixon 03]. All these are of type II$(3, 3, 1)$.

Theorem 9.1 (Pfaff–Saalschütz Formula). *If* $a_1 + a_2 + b_1 + b_2 = n - 1$, *then*

$$\sum_{k=0}^{n} \binom{n}{k} [a_1]_k [a_2]_k [b_1]_{n-k} [b_2]_{n-k} (-1)^k = [a_1 + b_1]_n [a_1 + b_2]_n = [b_1 + a_1]_n [b_1 + a_2]_n (-1)^n. \tag{9.1}$$

The quotient is

$$q_{ps} = \frac{(n-k)(a_1-k)(a_2-k)}{(-1-k)(n-1-b_1-k)(n-1-b_2-k)}.$$

Remark 9.2. The formula is reflexive. If we reverse the order of summation, the condition repeats itself. But the sign will change for odd n.

Theorem 9.3 (Symmetric Dixon Formula).

$$\sum_{k=0}^{n} \binom{n}{k} [a]_k [b]_k [a]_{n-k} [b]_{n-k} (-1)^k = \begin{cases} 0 & \text{for } n \text{ odd}, \\ [a]_m [b]_m [n-a-b-1]_m [n]_m & \text{for } n = 2m. \end{cases} \tag{9.2}$$

Theorem 9.4 (Balanced Dixon Formula).

$$\sum_{k=0}^{n}\binom{n}{k}[n+2a]_k[b+a]_k[b-a]_{n-k}[n-2a]_{n-k}(-1)^k$$
$$= [n-2a-1,2]_n[n+2b,2]_n. \qquad (9.3)$$

Besides these, there are formulas due to G. N. Watson [Watson 25], and F. J. W. Whipple [Whipple 25] in 1925, but both are transformations of the above formulas due to Dixon.

9.2 Transformations of Sums of Type II(3,3,1)

The sums of form

$$S_n(a_1,a_2,b_1,b_2)=\sum_{k=0}^{n}\binom{n}{k}[a_1]_k[a_2]_k[b_1]_{n-k}[b_2]_{n-k}(-1)^k \qquad (9.4)$$

are independent of interchanging a_1 and a_2 or b_1 and b_2, and of reversing the direction of summation, except for the factor $(-1)^n$, which depends on the interchange of the pair of sets $\{a_1,a_2\}$ and $\{b_1,b_2\}$.

Let us define the function of two variables

$$f_n(x,y)=n-1-x-y$$

and the transformation of the set of four variables

$$\tau_n(a_1,a_2,b_1,b_2)=(f_n(a_1,b_1),b_2,a_1,f_n(a_2,b_2)). \qquad (9.5)$$

With these notations we have

Theorem 9.5 (Transformation Theorem). *For S defined by (9.4) and the transformation of indices τ_n defined by (9.5), we have*

$$S_n(\tau_n(a_1,a_2,b_1,b_2))=S_n(a_1,a_2,b_1,b_2), \qquad (9.6)$$

or, written out,

$$S_n(n-1-a_1-b_1,b_2,a_1,n-1-a_2-b_2)=S_n(a_1,a_2,b_1,b_2). \qquad (9.7)$$

Proof: By Chu–Vandermonde (8.2), using (2.1), we have

$$\begin{aligned}[b_1]_{n-k}&=(-1)^{n-k}[-b_1+n-k-1]_{n-k}\\&=(-1)^{n-k}\sum_{j=k}^{n}\binom{n-k}{j-k}[a_1-k]_{j-k}[n-a_1-b_1-1]_{n-j}.\end{aligned}$$

Applied to the definition (9.4), it gives the form

$$S_n(a_1, a_2, b_1, b_2) = \sum_{k=0}^{n}\sum_{j=k}^{n}\binom{n}{k}\binom{n-k}{j-k}[a_1]_k[a_1-k]_{j-k}[a_2]_k \times [n-a_1-b_1-1]_{n-j}[b_2]_{n-k}(-1)^n.$$

Now we apply (2.9) and (2.2), and interchange the order of summation:

$$S_n(a_1, a_2, b_1, b_2) = (-1)^n\sum_{j=0}^{n}\binom{n}{j}[a_1]_j[n-a_1-b_1-1]_{n-j}\sum_{k=0}^{j}\binom{j}{k}[a_2]_k[b_2]_{n-k}.$$

Then we apply (2.1) to write $[b_2]_{n-k} = [b_2]_{n-j}[b_2-n+j]_{j-k}$, so we get

$$S_n(a_1, a_2, b_1, b_2) = (-1)^n\sum_{j=0}^{n}\binom{n}{j}[a_1]_j[n-a_1-b_1-1]_{n-j}[b_2]_{n-j} \times \sum_{k=0}^{j}\binom{j}{k}[a_2]_k[b_2-n+j]_{j-k}.$$

Now we can apply Chu–Vandermonde (8.2) to the inner sum and get

$$S_n(a_1, a_2, b_1, b_2) = (-1)^n\sum_{j=0}^{n}\binom{n}{j}[a_1]_j[n-a_1-b_1-1]_{n-j}[b_2]_{n-j}[a_2+b_2-n+j]_j.$$

When we apply (2.1) to the last term, we get

$$[a_2+b_2-n+j]_j = (-1)^j[n-a_2-b_2-1]_j,$$

and the proof is finished after changing the direction of summation. □

A simple consequence of the transformation theorem is the following formula.

Theorem 9.6. *For any $a, b, c \in \mathbb{C}$ and $n, p \in \mathbb{N}_0$ we have*

$$\sum_{k=0}^{n}\binom{n}{k}[a]_k[b]_{n-k}[c-k]_p = [a+b-p]_{n-p}\sum_{j=0}^{p}\binom{p}{j}[n]_j[b]_j[c-n]_{p-j}[p-1-a-b]_{p-j}(-1)^{p-j}.$$

Proof: We apply Lemma 5.11 to formula (4.25) to write

$$[c-k]_p = [c-p]_k [n-1-c]_{n-k} (-1)^k \frac{(-1)^n}{[c-p]_{n-p}}.$$

Then the transformation (9.6) yields

$$[a+b-p]_{n-p} \sum_{j=0}^{n} \binom{n}{j} [p]_j [b]_j [c-n]_{p-j} [p-1-a-b]_{p-j} (-1)^{p-j},$$

and we only have to note that $\binom{n}{j}[p]_j = \binom{p}{j}[n]_j$. □

By interchanging the a's or b's or both we can get this transformation in four different forms, all looking like

$$(a_1, a_2, b_1, b_2) \to \big(b, f_n(a, \hat{b}), a, f_n(\hat{a}, b)\big), \quad \hat{a} \neq a, \hat{b} \neq b.$$

Furthermore, by iteration of the transformation τ_n, one gets another four formulas, where we remark that

$$f_n\big(f_n(a,b), f_n(\hat{a}, \hat{b})\big) = a_1 + a_2 + b_1 + b_2 - n + 1$$

is independent of interchanges. Hence, we get the forms for $a \in \{a_1, a_2\}$ and $b \in \{b_1, b_2\}$,

$$(a_1, a_2, b_1, b_2) \to (a, a_1 + a_2 + b_1 + b_2 - n + 1, f_n(a, b_1), f_n(a, b_2)), \quad (9.8)$$

or

$$(a_1, a_2, b_1, b_2) \to (f_n(a_1, b), f_n(a_2, b), a_1 + a_2 + b_1 + b_2 - n + 1, b). \quad (9.9)$$

Not only is the Pfaff–Saalschütz theorem a simple consequence of the transformations (9.8) or (9.9), but they give us a generalization:

Theorem 9.7 (Generalized Pfaff–Saalschütz Formula). *If*

$$p = a_1 + a_2 + b_1 + b_2 - n + 1 \in \mathbb{N}_0, \quad (9.10)$$

then we have

$$\begin{aligned}
&\sum_{k=0}^{n} \binom{n}{k} [a_1]_k [a_2]_k [b_1]_{n-k} [b_2]_{n-k} (-1)^k \\
&\quad = \sum_{k=0}^{p} \binom{p}{k} [n]_k [a_2]_k [a_1 + b_1 - p]_{n-k} [a_1 + b_2 - p]_{n-k} (-1)^k \\
&\quad = [a_1 + b_1 - p]_{n-p} [a_1 + b_2 - p]_{n-p} \\
&\quad\quad \times \sum_{k=0}^{p} \binom{p}{k} [n]_k [a_2]_k [a_1 + b_1 - n]_{p-k} [a_1 + b_2 - n]_{p-k} (-1)^k. \quad (9.11)
\end{aligned}$$

Proof: From (9.8) we get immediately

$$\sum_{k=0}^{n}\binom{n}{k}[a_1]_k[a_2]_k[b_1]_{n-k}[b_2]_{n-k}(-1)^k$$
$$=\sum_{k=0}^{n}\binom{n}{k}[p]_k[a_2]_k[n-1-a_2-b_1]_{n-k}[n-1-a_2-b_2]_{n-k}(-1)^k.$$

Now we apply the condition (9.11) to write $n-1-a_2-b_2=a_1+b_1-p$, and the formula (2.10) to write $\binom{n}{k}[p]_k=\binom{p}{k}[n]_k$. Then we use (2.2) to write $[a+b-p]_{n-k}=[a+b-p]_{n-p}[a+b-n]_{p-k}$. The limit of summation can be changed to p because of the binomial coefficient. This establishes (9.11). □

A special case of this formula for $p=1$ appears in [Slater 66, (III.16)].

We do not know of any general formulas when p of (9.10) is a negative integer, but the special case of $p=-2$ has a two argument family of formulas found by I. M. Gessel and D. Stanton as a generalization of a one argument family due to R. W. Gosper; see [Gessel and Stanton 82, (1.9)].

$$\sum_{k=0}^{n}\binom{n}{k}[a-b-ab]_k[-na-ab-1]_k[ab-a-1]_{n-k}$$
$$\times[n+na+b+ab-1]_{n-k}(-1)^k$$
$$=(n+1)[n+b-1]_n[-na-a-1]_n. \tag{9.12}$$

Formula (9.12) has excess $p=-2$ (9.10). This is a transformation by τ_n (9.5) of the formula mentioned, which looks in our notation like

$$\sum_{k=0}^{n}\binom{n}{k}[ab-a-1]_k[1-b]_k[n+b]_{n-k}[-na-ab-1]_{n-k}(-1)^k$$
$$=(n+1)[n+b-1]_n[-na-a-1]_n. \tag{9.13}$$

Proof: From (2.2) and (2.1) we have

$$[n+b]_{n+2}=[n+b]_{n-k}[b+k]_2[1-b]_k(-1)^k.$$

Dividing both sides of (9.13) by this factor, we get the equivalent identity

$$\sum_{k=0}^{n}\binom{n}{k}\frac{[ab-a-1]_k[-na-ab-1]_{n-k}}{[b+k]_2}=\frac{(n+1)[-na-a-1]_n}{(n+b)(b-1)}.$$

Splitting the denominator as $\frac{1}{[b+k]_2} = \frac{1}{b+k-1} - \frac{1}{b+k}$, we can split the sum in two, change the summation variable in one of them and join them again:

$$\begin{aligned}
&\sum_{k=0}^{n} \binom{n}{k} \frac{[ab-a-1]_k[-na-ab-1]_{n-k}}{[b+k]_2} \\
&= \sum_{k=0}^{n} \binom{n}{k} \frac{[ab-a-1]_k[-na-ab-1]_{n-k}}{b+k-1} \\
&\quad - \sum_{k=0}^{n} \binom{n}{k} \frac{[ab-a-1]_k[-na-ab-1]_{n-k}}{b+k} \\
&= \sum_{k=0}^{n} \binom{n}{k} \frac{[ab-a-1]_k[-na-ab-1]_{n-k}}{b+k-1} \\
&\quad - \sum_{k=1}^{n+1} \binom{n}{k-1} \frac{[ab-a-1]_{k-1}[-na-ab-1]_{n+1-k}}{b+k-1} \\
&= \sum_{k=1}^{n+1} \binom{n}{k-1} \frac{[ab-a-1]_{k-1}[-na-ab-1]_{n-k}}{b+k-1} \cdot \frac{(n+1)a(b+k-1)}{k} \\
&\quad + \frac{[-na-ab-1]_n}{b-1} \\
&= a \sum_{k=1}^{n+1} \binom{n+1}{k} [ab-a-1]_{k-1}[-na-ab-1]_{n-k} \\
&\quad + a[ab-a-1]_{-1}[-na-ab-1]_n \\
&= \frac{-1}{a(b-1)(n+b)} \sum_{k=0}^{n+1} \binom{n+1}{k} [ab-a]_k[-na-ab]_{n+1-k}.
\end{aligned}$$

Now we can apply the Chu–Vandermonde formula (8.2) to obtain

$$\frac{-1}{a(b-1)(n+b)} \cdot [-na-a]_{n+1} = \frac{(n+1)[-na-a-1]_n}{(b-1)(n+b)},$$

as desired. □

9.3 Generalizations of Dixon's Formulas

We want to consider a family of sums similar to the symmetric Dixon formula. So let us define for $n \in \mathbb{N}_0$ and $p, q \in \mathbb{Z}$

$$S_n(a+p, b+q, a, b) = \sum_{k=0}^{n} \binom{n}{k} [a+p]_k[b+q]_k[a]_{n-k}[b]_{n-k}(-1)^k. \quad (9.14)$$

For $p = q = 0$ we get the sum in the symmetric Dixon formula (9.2). We can find expressions for such sums for $p = q$.

Theorem 9.8 (Quasi-Symmetric Dixon Formulas). *For* $p \in \mathbb{N}_0$,

$$S_n(a+p, b+p, a, b) = \sum_{j=\lceil \frac{n}{2} \rceil}^{\lfloor \frac{n+p}{2} \rfloor} \binom{p}{2j-n} [n]_j \, [n-p-a-b-1]_j \, [a]_{n-j} \, [b]_{n-j} \, , \tag{9.15}$$

and

$$S_n(a-p, b-p, a, b) = (-1)^n \sum_{j=\lceil \frac{n}{2} \rceil}^{\lfloor \frac{n+p}{2} \rfloor} \binom{p}{2j-n} [n]_j \times [n+p-a-b-1]_j \, [a-p]_{n-j} \, [b-p]_{n-j} \, . \tag{9.16}$$

Proof: To prove (9.15), the Pfaff–Saalschütz formula (9.1) yields

$$[a+p]_k [b+p]_k = \sum_{j=0}^{k} \binom{k}{j} [\alpha]_j [\beta]_j [a+p-\alpha]_{k-j} [b+p-\alpha]_{k-j} (-1)^j, \tag{9.17}$$

provided $\beta = \alpha + k - 2p - a - b - 1$. When we substitute (9.17) into the sum (9.14) with $q = p$, we get

$$S_n(a+p, b+p, a, b) = \sum_{k=0}^{n} \sum_{j=0}^{k} \binom{n}{k} \binom{k}{j} [\alpha]_j [\beta]_j [a+p-\alpha]_{k-j} \times [b+p-\alpha]_{k-j} [a]_{n-k} [b]_{n-k} (-1)^{j+k}.$$

Now we apply (2.9) to write $\binom{n}{k}\binom{k}{j} = \binom{n}{j}\binom{n-j}{k-j}$. Then choose $\alpha = n+p-k$, so that we can apply (2.2) to write $[a]_{n-k}[a+p-\alpha]_{k-j} = [a]_{n-j}$ and $[b]_{n-k}[b+p-\alpha]_{k-j} = [b]_{n-j}$. Remembering that $\beta = n-p-a-b-1$, eventually we interchange the order of summations to obtain

$$S_n(a+p, b+p, a, b) = \sum_{j=0}^{n} \binom{n}{j} [a]_{n-j} [b]_{n-j} [n-p-a-b-1]_j \times \sum_{k=j}^{n} \binom{n-j}{k-j} [n+p-k]_j (-1)^{j+k}.$$

When we apply (8.9) to the inner sum, it can be evaluated as

$$[n+p-j-n+j]_{j-n+j} [j]_{n-j} = [p]_{2j-n} [j]_{n-j}.$$

We only need to write $\binom{n}{j}[p]_{2j-n}[j]_{n-j} = \binom{p}{2j-n}[n]_j$ and restrict summation to nonzero terms to obtain (9.15).

To prove (9.16) we reverse the direction of summation and apply (9.15) to $(a-p, b-p)$. □

We do not know formulas for the general sums (9.14), but we can furnish a couple of recursion formulas.

Theorem 9.9 (General Recursion Formula). *The sums in* (9.14) *satisfy*

$$\begin{aligned} S_n(a+p, b+q, a, b) &= S_n(a+p-1, b+q, a, b) \\ &\quad - n(b+q)S_{n-1}(a+p-1, b+q-1, a, b). \end{aligned} \tag{9.18}$$

Proof: We apply (2.4) to write $[a+p]_k = [a+p-1]_k + k[a+p-1]_{k-1}$. Then the sum splits into the two mentioned in (9.18). □

The interesting particular case of $q = 0$ can be treated by a recursion of such sums:

Theorem 9.10 (Special Recursion Formula). *The sums in* (9.14) *satisfy*

$$\begin{aligned} S_n(a+p, b, a, b) &= S_n(a+p-1, b, a, b) \\ &\quad - n(b-a-p+1)S_{n-1}(a+p-1, b, a, b) \\ &\quad - n(a+p-1)S_{n-1}(a+p-2, b, a, b). \end{aligned} \tag{9.19}$$

Proof: We apply (2.4) to write $[a+p]_k = [a+p-1]_k + k[a+p-1]_{k-1}$. Then the sum splits into two. The first one is just $S_n(a+p-1, b, a, b)$, but the second is

$$\begin{aligned} S_n(a+p, b, a, b) &= S_n(a+p-1, b, a, b) \\ &\quad + \sum_{k=0}^{n} \binom{n}{k} k[a+p-1]_{k-1}[b]_k[a]_{n-k}[b]_{n-k}(-1)^k \\ &= S_n(a+p-1, b, a, b) \\ &\quad + n\sum_{k=1}^{n} \binom{n-1}{k-1} [a+p-1]_{k-1}[b]_k[a]_{n-k}[b]_{n-k}(-1)^k \\ &= S_n(a+p-1, b, a, b) - nT_{n-1}(a, b, p-1), \end{aligned} \tag{9.20}$$

where we have put

$$T_n(a, b, p) = \sum_{k=0}^{n} \binom{n}{k} [a+p]_k[b]_{k+1}[a]_{n-k}[b]_{n-k}(-1)^k.$$

If we write $[b]_{k+1} = [b]_k(b-k)$, then we get the new sum T_n, written as a sum of two,

$$\begin{aligned} T_n(a,b,p) &= bS_n(a+p,b,a,b) \\ &\quad + n\sum_{k=0}^{n-1}\binom{n-1}{k}[a+p]_{k+1}[b]_{k+1}[a]_{n-1-k}[b]_{n-1-k}(-1)^k \\ &= bS_n(a+p,b,a,b) + n(a+p)T_{n-1}(a,b,p-1). \end{aligned} \tag{9.21}$$

If we multiply (9.20) by $a+p$ and add to (9.21), then we get

$$T_n(a,b,p) = (b-a-p)S_n(a+p,b,a,b) + (a+p)S_n(a+p-1,b,a,b). \tag{9.22}$$

When we substitute (9.22) into (9.20), we eventually arrive at the recursion (9.19). □

To get started we need two neighboring values, so we need to compute $S_n(a+1,b,a,b)$.

From (9.18) we get

$$S_n(a+1,b,a,b) = S_n(a+1,b+1,a,b) + n(a+1)S_{n-1}(a,b,a,b),$$

where the two terms follow from (9.15). The result becomes

$$S_n(a+1,b,a,b) = [n]_{\lfloor \frac{n}{2} \rfloor}[n-a-b-2]_{\lceil \frac{n}{2} \rceil}[a]_{\lceil \frac{n}{2} \rceil}[b]_{\lfloor \frac{n}{2} \rfloor}(-1)^n. \tag{9.23}$$

9.4 The Balanced and Quasi-Balanced Dixon Identities

The balanced Dixon identity (9.3) is not so easy to generalize. And its proof does not follow from the Pfaff–Saalschütz formula or the symmetric Dixon theorem. We will present a proof following an idea due to G. N. Watson [Watson 24].

To also prove some quasi-balanced Dixon identities, we will consider small integers $p \in \mathbb{Z}$, with the balanced case as $p = 0$,

$$\begin{aligned} &S_n(n+2a, b+a, b-a, n-2a-p) \\ &\quad = \sum_{k=0}^{n}\binom{n}{k}[n+2a]_k[b+a]_k[b-a]_{n-k}[n-2a-p]_{n-k}(-1)^k. \end{aligned}$$

Now we apply the Chu–Vandermonde formula (8.3) to write

$$[a+b]_k = \sum_{j=0}^{k} \binom{k}{j} [n+2a-k]_j [b-a-n+k]_{k-j}.$$

Then the sum becomes

$$\begin{aligned} &S_n(n+2a, b+a, b-a, n-2a-p) \\ &= \sum_{k=0}^{n}\sum_{j=0}^{k} \binom{n}{k}\binom{k}{j} [n+2a]_k [n+2a-k]_j \\ &\quad \times [b-a]_{n-k}[b-a-n+k]_{k-j}[n-2a-p]_{n-k}(-1)^k. \end{aligned}$$

Now we apply (2.9) to write $\binom{n}{k}\binom{k}{j} = \binom{n}{j}\binom{n-j}{k-j}$. Next apply (2.2) twice and change the order of summations to get

$$\begin{aligned} &S_n(n+2a, b+a, b-a, n-2a-p) \\ &= \sum_{j=0}^{n} \binom{n}{j} [b-a]_{n-j} \sum_{k=j}^{n} \binom{n-j}{k-j} [n+2a]_{k+j}[n-2a-p]_{n-k}(-1)^k. \end{aligned}$$

Now we apply (2.2) to write $[n+2a]_{k+j} = [n+2a]_{2j}[n+2a-2j]_{k-j}$. Then the sum becomes

$$\begin{aligned} &S_n(n+2a, b+a, b-a, n-2a-p) \\ &= \sum_{j=0}^{n} \binom{n}{j} [n+2a]_{2j}[b-a]_{n-j} \\ &\quad \times \sum_{k=j}^{n} \binom{n-j}{k-j} [n+2a-2j]_{k-j}[n-2a-p]_{n-k}(-1)^k. \end{aligned} \tag{9.24}$$

Changing variables in the inner sum, one gets

$$(-1)^j \sum_{k=0}^{n-j} \binom{n-j}{k} [n-j-p+(2a-j+p)]_k [n-j-(2a-j+p)]_{n-j-k}(-1)^k.$$

This sum is a quasi-balanced Kummer sum, to be evaluated by (8.20) as

$$\begin{aligned} &\frac{2^{n-j-(p\vee 0)}}{[n-j-p]_{-(p\wedge 0)}} \sum_{i=0}^{|p|} \binom{|p|}{i} \sigma(p)^{p+i} \\ &\qquad\qquad \times [-(2a-j+p)+n-j+i-1, 2]_{n-j-(p\wedge 0)} \\ &= \frac{2^{n-j-(p\vee 0)}}{[n-j-p]_{-(p\wedge 0)}} \sum_{i=0}^{|p|} \binom{|p|}{i} \sigma(p)^{p+i} [-2a+p+n+i-1, 2]_{n-j-(p\wedge 0)}. \end{aligned} \tag{9.25}$$

If $p = 0$, we apply (2.1) to (9.25) and substitute the result into (9.24) to get

$$S_n(n+2a, b+a, b-a, n-2a)$$
$$= \sum_{j=0}^{n} \binom{n}{j} [n+2a]_{2j} [b-a]_{n-j} (-1)^j (-2)^{n-j} [n+2a-2j-1, 2]_{n-j}.$$

If we split $[n+2a]_{2j} = [n+2a, 2]_j [n+2a-1, 2]_j$ and use (2.4) to get

$$[n+2a-1, 2]_j [n+2a-2j-1, 2]_{n-j} = [n+2a-1, 2]_n$$

and (2.5) to get

$$[b-a]_{n-j} 2^{n-j} = [2b-2a, 2]_{n-j},$$

we can write the sum as

$$S_n(n+2a, b+a, b-a, n-2a)$$
$$= [n+2a-1, 2]_n (-1)^n \sum_{j=0}^{n} \binom{n}{j} [n+2a, 2]_j [2b-2a, 2]_{n-j}.$$

Now the Chu–Vandermonde formula (8.3) applies, so we get

$$S_n(n+2a, b+a, b-a, n-2a) = (-1)^n [n+2a-1, 2]_n [n+2b, 2]_n.$$

This proves the balanced Dixon identity (9.3).

If $p > 0$, we get

$$S_n(n+2a, b+a, b-a, n-2a-p)$$
$$= (-1)^n 2^{-p} \sum_{i=0}^{p} \binom{p}{i} \sum_{j=0}^{n} \binom{n}{j} [2b-2a, 2]_{n-j} [n+2a, 2]_j$$
$$\times [n+2a-1, 2]_j [n+2a-2j+p-1-i, 2]_{n-j}, \quad (9.26)$$

and for $p < 0$ we get

$$S_n(n+2a, b+a, b-a, n-2a-p) = \frac{(-1)^n}{[n-p]_{-p} [2b-2a-2p, 2]_{-p}}$$
$$\times \sum_{i=0}^{-p} \binom{-p}{i} (-1)^i \sum_{j=0}^{n} \binom{n-p}{j} [2b-2a-2p, 2]_{n-p-j} [n+2a, 2]_j$$
$$\times [n+2a-1, 2]_j [n+2a-2j-p-1-i, 2]_{n-p-j}.$$

(9.27)

Theorem 9.11. *Some quasi-balanced Dixon formulas are*

$$\begin{aligned}
&S_n(n+2a, b+a, b-a, n-2a-1)\\
&\quad = \tfrac{1}{2}\big([n-2b-1,2]_n[n+2a,2]_n\\
&\qquad + [n-2a-1,2]_n[n+2b,2]_n\big),
\end{aligned}\tag{9.28}$$

$$\begin{aligned}
&S_n(n+2a, b+a, b-a, n-2a-2)\\
&\quad = \tfrac{1}{4}[n-2a-1,2]_n[n+2b,2]_n\\
&\qquad + \tfrac{1}{2}[n-2b-1,2]_n[n+2a,2]_n\\
&\qquad + \tfrac{1}{4}[n-2a-3,2]_n[n+2b,2]_n\\
&\qquad + \tfrac{1}{2}n(n+2a)[n-2b-2,2]_n[n+2a-1,2]_n,
\end{aligned}\tag{9.29}$$

$$\begin{aligned}
&S_n(n+2a, b+a, b-a, n-2a+1) = \frac{(-1)^n}{(n+1)(2b-2a+2)}\\
&\quad \times \Big([n+2a,2]_{n+1}[n+2b+1,2]_{n+1}\\
&\qquad - [n+2a-1,2]_{n+1}[n+2b+2,2]_{n+1}\Big).
\end{aligned}\tag{9.30}$$

Proof: Substitute $p=1$ and $p=2$ in (9.26) and $p=-1$ in (9.27) and eventually apply the Chu–Vandermonde formula (8.3) to each term. ☐

Remarks 9.12. Surprisingly, $S_n(n+2a, b+a, b-a, n-2a-1)$ is symmetric in (a,b). Another symmetry for the balanced Dixon formula is

$$\begin{aligned}
&S_n(n+2a, b+a, b-a, n-2a)\\
&\qquad = S_n\left(n-2b-1, -b-a-1, b-a, n+2b+1\right).
\end{aligned}\tag{9.31}$$

Furthermore, it does not matter which term is changed by the deviation p. If we apply (9.6) to $S_n(n+2a, b+a, b-a, n-2a-p)$, we get the symmetric expression

$$\sum_{k=0}^{n}\binom{n}{k}[-a-b-1]_k[a+b]_k[b-a]_{n-k}[-1+a-b+p]_{n-k}(-1)^k.$$

Proof of (9.31)*:* Apply the transformation (9.6). ☐

9.5 Watson's Formulas and Their Contiguous Companions

The two finite versions of Watson's formulas are

$$\sum_{k=0}^{n}\binom{n}{k}[-a]_k[-b]_k\left[\tfrac{1}{2}(n+a-1)\right]_{n-k}[n+2b-1]_{n-k}(-1)^k$$
$$=\begin{cases}0 & \text{for } n \text{ odd},\\ \left[-\tfrac{1}{2}(a+1)\right]_m[-b]_m\left[\tfrac{1}{2}(a-1)-b\right]_m[n]_m & \text{for } n=2m,\end{cases} \tag{9.32}$$

$$\sum_{k=0}^{n}\binom{n}{k}[-a]_k[-b]_k\left[n+\tfrac{1}{2}(a+b-1)\right]_{n-k}[-n-1]_{n-k}(-1)^k$$
$$=[-a-1,2]_n[-b-1,2]_n(-1)^n. \tag{9.33}$$

In addition, we can introduce an excess p in one or two of the factorials. We have:

$$\sum_{k=0}^{n}\binom{n}{k}[-a]_k[-b]_k\left[\tfrac{1}{2}(n+a-1)\right]_{n-k}[n+2b]_{n-k}(-1)^k \tag{9.34}$$
$$=[n]_{\lfloor\frac{n}{2}\rfloor}\left[\tfrac{1}{2}(n-a-1)+b\right]_{\lceil\frac{n}{2}\rceil}[-b-1]_{\lceil\frac{n}{2}\rceil}\left[\tfrac{1}{2}(n+a-1)\right]_{\lfloor\frac{n}{2}\rfloor}(-1)^n,$$

$$\sum_{k=0}^{n}\binom{n}{k}[-a]_k[-b]_k\left[\tfrac{1}{2}(n+a-1)\right]_{n-k}[n+2b-2]_{n-k}(-1)^k$$
$$=[n]_{\lfloor\frac{n}{2}\rfloor}\left[\tfrac{1}{2}(n-a-3)+b\right]_{\lceil\frac{n}{2}\rceil}[-b]_{\lceil\frac{n}{2}\rceil}\left[\tfrac{1}{2}(n+a-1)\right]_{\lfloor\frac{n}{2}\rfloor}, \tag{9.35}$$

$$\sum_{k=0}^{n}\binom{n}{k}[-a-p]_k[-b]_k\left[\tfrac{1}{2}(n+a-1)\right]_{n-k}[n+2b-1+p]_{n-k}(-1)^k$$
$$=\sigma(p)^n\sum_{j=\lceil\frac{n}{2}\rceil}^{\lfloor\frac{n+|p|}{2}\rfloor}\binom{|p|}{2j-n}[n]_j\left[\tfrac{1}{2}(n-a-1)+b\right]_j$$
$$\times\left[\tfrac{1}{2}(n+a-1)+(p\wedge 0)\right]_{n-j}[-b-(p\vee 0)]_{n-j}, \tag{9.36}$$

$$\sum_{k=0}^{n}\binom{n}{k}[-a]_k[-b]_k\left[n+\tfrac{1}{2}(a+b-1)\right]_{n-k}[-n]_{n-k}(-1)^k$$
$$=\tfrac{1}{2}\left([-a,2]_n[-b,2]_n+[-a-1,2]_n[-b-1,2]_n\right)(-1)^n, \tag{9.37}$$

$$\sum_{k=0}^{n}\binom{n}{k}[-a]_k[-b]_k\left[n+\tfrac{1}{2}(a+b-1)\right]_{n-k}[-n-2]_{n-k}(-1)^k$$
$$=\frac{1}{(n+1)(2n+a+b+1)}\big([-a,2]_{n+1}[-b,2]_{n+1}$$
$$+[-a-1,2]_{n+1}[-b-1,2]_{n+1}\big)(-1)^n. \tag{9.38}$$

Proof: Formulas (9.32) and (9.34)–(9.36) are transformed by (9.9) into the different quasi-symmetric Dixon formulas, (9.2), (9.23), and (9.15)–(9.16). Formulas (9.33) and (9.37)–(9.38) are similarly transformed by (9.9) into the balanced Dixon formula (9.3) and the quasi-balanced Dixon formulas (9.28) and (9.30). □

Formulas (9.34), (9.35), (9.37), and (9.38), with p of size 1, were considered by J. L. Lavoie in 1987 [Lavoie 87].

9.6 Whipple's Formulas and Their Contiguous Companions

The two finite versions of Whipple's formulas are

$$\sum_{k=0}^{n}\binom{n}{k}[-n-1]_k[-a]_k[b+n-1]_{n-k}[2a-b+n]_{n-k}(-1)^k$$
$$=[b+n-2,2]_n[2a-b+n-1,2]_n, \tag{9.39}$$

$$\sum_{k=0}^{n}\binom{n}{k}[a-1]_k[-a]_k[b+n-1]_{n-k}[-b-n]_{n-k}(-1)^k$$
$$=[a-b-1,2]_n[-a-b,2]_n. \tag{9.40}$$

These also allow an excess of size 1, giving the following four formulas:

$$\sum_{k=0}^{n}\binom{n}{k}[-n-1]_k[-a]_k[b+n-1]_{n-k}[2a-b+n+1]_{n-k}(-1)^k$$
$$=\tfrac{1}{2}\Big([b+n-2,2]_n[2a-b+n-1,2]_n$$
$$[b+n-1,2]_n[2a-b+n-2,2]_n\Big), \tag{9.41}$$

$$\sum_{k=0}^{n}\binom{n}{k}[-n-1]_k[-a]_k[b+n-1]_{n-k}[2a-b+n-1]_{n-k}(-1)^k$$
$$=\frac{(-1)^n}{(n+1)2a}\Big([b+n-1,2]_{n+1}[-2a+b+n,2]_{n+1}$$
$$-[b+n-2,2]_{n+1}[2a-b+n-1,2]_{n+1}\Big), \tag{9.42}$$

$$\sum_{k=0}^{n}\binom{n}{k}[a-1]_k[-a]_k[b+n-1]_{n-k}[-b-n+1]_{n-k}(-1)^k$$
$$=\tfrac{1}{2}\left([a-b-1,2]_n[-a-b,2]_n+[a-b,2]_n[-a-b+1,2]_n\right), \tag{9.43}$$

$$\sum_{k=0}^{n}\binom{n}{k}[a-1]_k[-a]_k[b+n-1]_{n-k}[-b-n-1]_{n-k}(-1)^k$$
$$= \frac{1}{(n+1)2(n+b)}\Big([a-b-1,2]_{n+1}[-a-b+2,2]_{n+1}$$
$$- [a-b,2]_{n+1}[-a-b-1,2]_{n+1}\Big). \tag{9.44}$$

Proof: The formulas (9.39), (9.41), and (9.42) are transformed by (9.9) to the different quasi-balanced Dixon formulas, (9.3), (9.28), and (9.30). The formulas (9.40), (9.43), and (9.44) are respectively the balanced Dixon formula, (9.4) transformed by (9.6) and once more transformed by (9.6) to the quasi-balanced Dixon formulas, (9.28) and (9.30). □

9.7 Ma Xin-Rong and Wang Tian-Ming's Problem

In 1995 Ma Xin-Rong and Wang Tian-Ming [Ma and Wang 95] posed the following problem: *Show that*

$$\sum_{k=0}^{m}\binom{m}{k}\binom{n+k}{k}\binom{n+1}{j-k} = \sum_{k=0}^{n}\binom{n}{k}\binom{m+k}{k}\binom{m+1}{j-k}, \tag{9.45}$$

and for $m \leq n-1$,

$$\sum_{k=0}^{m}\binom{2n-m-1-k}{n-k}\binom{m+k}{k} = \binom{2n-1}{n}. \tag{9.46}$$

Proof of (9.45)*:* We remark that the upper limits of summation may well be changed to j, because possible terms with $k > j$ or $k > m$ $(k > n)$ are zeros anyway. Hence, the identity can be written as

$$\sum_{k=0}^{j}\binom{m}{k}\binom{n+k}{k}\binom{n+1}{j-k} = \sum_{k=0}^{j}\binom{n}{k}\binom{m+k}{k}\binom{m+1}{j-k}. \tag{9.47}$$

In the form (9.47), the sums are defined for any complex values of m and n.

We will prove the generalization of (9.45), that the function

$$S_j(m,n) = \sum_{k=0}^{j}\binom{m}{k}\binom{n+k}{k}\binom{n+1}{j-k}, \tag{9.48}$$

defined for integers j and complex variables m and n, is symmetric in m and n.

The function (9.48) can be multiplied by $j!^2$ without disturbing a possible symmetry. Then the function $j!^2 S_j(m,n)$ becomes, using the rewriting $[n+k]_k = [-n-1]_k(-1)^k$, cf. (2.1),

$$j!^2 S_j(m,n) = \sum_{k=0}^{j} \binom{j}{k} [m]_k[-n-1]_k[n+1]_{j-k}[j]_{j-k}(-1)^k. \tag{9.49}$$

Now we apply the transformation (9.7) to write (9.49) in a form symmetric in m and n

$$j!^2 S_j(m,n) = \sum_{k=0}^{j} \binom{j}{k} [m]_k[n]_k[j-2-n-m]_{j-k}[j]_{j-k}(-1)^{j-k}.$$

If we divide again by $j!^2$, where we use once more

$$[j-2-n-m]_{j-k}(-1)^{j-k} = [m+n+1-k]_{j-k},$$

then we get one symmetric form of the original function (9.48):

$$S_j(m,n) = \sum_{k=0}^{j} \binom{m}{k}\binom{n}{k}\binom{m+n+1-k}{j-k}.$$

If we apply (9.7) to the function (9.49), interchanging the first two factorials,

$$j!^2 S_j(m,n) = \sum_{k=0}^{j} \binom{j}{k} [-n-1]_k[m]_k[n+1]_{j-k}[j]_{j-k}(-1)^k,$$

then we get another symmetric form, namely

$$j!^2 S_j(m,n) = (-1)^j \sum_{k=0}^{j} \binom{j}{k} [-n-1]_k[-m-1]_k[j-1]_{j-k}[j]_{j-k}(-1)^k.$$

From this form we can regain as another symmetric form of the function (9.48),

$$S_j(m,n) = \sum_{k=0}^{j} \binom{-n-1}{k}\binom{-m-1}{k}\binom{j-1}{j-k}(-1)^{j-k},$$

or, if you prefer,

$$S_j(m,n) = \sum_{k=0}^{j} \binom{n+k}{k}\binom{m+k}{k}\binom{-k}{j-k}.$$

□

Proof of (9.46)*:* We remark that (9.46) is true for $m = 0$. Then we prove that as long as $m < n$, the sum is the same for m and $m-1$. To get this equality it is convenient to apply Abelian summation or summation by parts. We remark that by (2.7) we have

$$\binom{2n-m-1-k}{n-k} = \binom{2n-(m-1)-1-k}{n-k} - \binom{2n-(m-1)-1-k-1}{n-k-1}.$$

Using this we can split the sum in (9.46) as

$$\begin{aligned}
&\sum_{k=0}^{m} \binom{2n-m-1-k}{n-k}\binom{m+k}{k} \\
&= \sum_{k=0}^{m} \binom{2n-(m-1)-1-k}{n-k}\binom{m+k}{k} \\
&\qquad - \sum_{k=0}^{m} \binom{2n-(m-1)-1-k-1}{n-k-1}\binom{m+k}{k} \\
&= \sum_{k=0}^{m} \binom{2n-(m-1)-1-k}{n-k}\binom{m+k}{k} \\
&\qquad - \sum_{k=0}^{m+1} \binom{2n-(m-1)-1-k}{n-k}\binom{m+k-1}{k-1} \\
&= \sum_{k=0}^{m-1} \binom{2n-(m-1)-1-k}{n-k}\binom{(m-1)+k}{k} \\
&\qquad + \binom{2n-2m}{n-m}\binom{2m-1}{m} - \binom{2n-2m-1}{n-m-1}\binom{2m}{m},
\end{aligned} \tag{9.50}$$

where the last two terms cancel as long as $m < n$, because one is obtained from the other by moving a factor 2 from one binomial coefficient to the other, cf. (5.11).

If we have $m = n$, then the equation (9.50) becomes

$$\binom{2n}{n} = \binom{2n-1}{n} + \binom{2n-1}{n} - 0,$$

where the last two terms do not cancel. $\square$

Comment. If we replace m with $n-m-1$ in the formula (9.46), then we get

$$\sum_{k=0}^{n-m-1} \binom{n+m-k}{n-k}\binom{n-m-1+k}{k} = \binom{2n-1}{n} \quad (m \leq n-1).$$

Now, if we change the summation variable, replacing k with $n-k$, i.e., changing the direction of summation, then we get

$$\sum_{k=m+1}^{n} \binom{m+k}{k}\binom{2n-m-1-k}{n-k} = \binom{2n-1}{n} \quad (m \leq n-1). \qquad (9.51)$$

But this is the natural continuation of (9.46), so if we add (9.46) and (9.51), we get

$$\sum_{k=0}^{n} \binom{2n-m-1-k}{n-k}\binom{m+k}{k} = \binom{2n}{n},$$

which happens to be valid for all complex values of m.

This is just a simple consequence of the Chu–Vandermonde equation (8.3). To see this we only need to change the binomial coefficients by (2.11):

$$\begin{aligned}
&\sum_{k=0}^{n} \binom{2n-m-1-k}{n-k}\binom{m+k}{k} \\
&\quad = (-1)^n \sum_{k=0}^{n} \binom{n-k-2n+m+1+k-1}{n-k}\binom{k-m-k-1}{k} \\
&\quad = (-1)^n \sum_{k=0}^{n} \binom{m-n}{n-k}\binom{-m-1}{k} \\
&\quad = (-1)^n \binom{-n-1}{n} \\
&\quad = \binom{2n}{n}.
\end{aligned}$$

9.8 C. C. Grosjean's Problem

In 1992, C. C. Grosjean [Grosjean 92, Andersen and Larsen 93] posed the following problem:

Determine the sum

$$\sum_{m=0}^{n} \binom{-\frac{1}{4}}{m}^2 \binom{-\frac{1}{4}}{n-m}^2.$$

Solution. Consider the sum

$$S = \sum_{m=0}^{n} \binom{n}{m} [n+x+y]_m \left[x-\tfrac{1}{4}\right]_m \left[y-\tfrac{1}{4}\right]_m \times [n-x-y]_{n-m} \left[-x-\tfrac{1}{4}\right]_{n-m} \left[-y-\tfrac{1}{4}\right]_{n-m}. \tag{9.52}$$

We use the Pfaff–Saalschütz identity (9.1),

$$[a+d]_n [b+d]_n = \sum_{k=0}^{n} \binom{n}{k} [n-a-b-d-1]_k [d]_k [a]_{n-k} [b]_{n-k} (-1)^k. \tag{9.53}$$

We want to write the sum in (9.52) as a double sum using (9.53) to expand $\left[x-\frac{1}{4}\right]_m \left[y-\frac{1}{4}\right]_m$. We therefore replace n by m and let $a = x - \frac{1}{4} - d$, $b = y - \frac{1}{4} - d$ in (9.53). This gives

$$S = \sum_{m=0}^{n} \binom{n}{m} [n+x+y]_m [n-x-y]_{n-m} \left[-x-\tfrac{1}{4}\right]_{n-m} \left[-y-\tfrac{1}{4}\right]_{n-m}$$
$$\times \sum_{k=0}^{m} \binom{m}{k} \left[m+d-x-y-\tfrac{1}{2}\right]_k [d]_k \left[x-\tfrac{1}{4}-d\right]_{m-k} \left[y-\tfrac{1}{4}-d\right]_{m-k} (-1)^k.$$

If we let $d = n - m + x + y$, then

$$\left[x-\tfrac{1}{4}-d\right]_{m-k} \left[-y-\tfrac{1}{4}\right]_{n-m} = \left[-y-\tfrac{1}{4}\right]_{n-k},$$

and

$$\left[y-\tfrac{1}{4}-d\right]_{m-k} \left[-x-\tfrac{1}{4}\right]_{n-m} = \left[-x-\tfrac{1}{4}\right]_{n-k}.$$

Using also $\binom{n}{m}\binom{m}{k} = \binom{n}{k}\binom{n-k}{m-k}$ and changing the order of summation, we obtain

$$S = \sum_{k=0}^{n} \binom{n}{k} \left[n-\tfrac{1}{2}\right]_k \left[-x-\tfrac{1}{4}\right]_{n-k} \left[-y-\tfrac{1}{4}\right]_{n-k} (-1)^k \times \sum_{m=k}^{n} \binom{n-k}{m-k} [n+x+y]_m [n-x-y]_{n-m} [n-m+x+y]_k. \tag{9.54}$$

Let T be the inner sum, i.e.,

$$T = \sum_{m=k}^{n} \binom{n-k}{m-k} [n+x+y]_m [n-x-y]_{n-m} [n-m+x+y]_k.$$

Since $[n+x+y]_m\,[n-m+x+y]_k = [n+x+y]_{2k}\,[n+x+y-2k]_{m-k}$, it follows using the Chu–Vandermonde convolution (8.3) and (5.12) that

$$T = [n+x+y]_{2k}\,[2n-2k]_{n-k} = \left[\tfrac{n+x+y}{2}\right]_k \left[\tfrac{n+x+y-1}{2}\right]_k \left[n-k-\tfrac{1}{2}\right]_{n-k} 4^n.$$

Introducing this into (9.54) yields

$$\begin{aligned} S &= \sum_{k=0}^{n} \binom{n}{k} \left[n-\tfrac{1}{2}\right]_k \left[-x-\tfrac{1}{4}\right]_{n-k} \left[-y-\tfrac{1}{4}\right]_{n-k} (-1)^k \\ &\qquad \times \left[\tfrac{n+x+y}{2}\right]_k \left[\tfrac{n+x+y-1}{2}\right]_k \left[n-k-\tfrac{1}{2}\right]_{n-k} 4^n \\ &= 4^n \left[n-\tfrac{1}{2}\right]_n \sum_{k=0}^{n} \binom{n}{k} \left[\tfrac{n+x+y}{2}\right]_k \left[\tfrac{n+x+y-1}{2}\right]_k \\ &\qquad \times \left[-x-\tfrac{1}{4}\right]_{n-k} \left[-y-\tfrac{1}{4}\right]_{n-k} (-1)^k. \end{aligned}$$

This is a Pfaff–Saalschütz sum (9.1), and using (9.53) we get an identity for the sum in (9.52):

$$S = [2n]_n \left[\tfrac{n-x+y}{2}-\tfrac{3}{4}\right]_n \left[\tfrac{n+x-y}{2}-\tfrac{3}{4}\right]_n = [2n]_n \left[n+x-y-\tfrac{1}{2}\right]_{2n} \left(-\tfrac{1}{4}\right)^n,$$

where we have given two different forms of the result. If we divide by $n!^3$ we obtain

$$\begin{aligned} &\sum_{m=0}^{n} \binom{n+x+y}{m} \binom{x-\frac{1}{4}}{m} \binom{y-\frac{1}{4}}{m} \\ &\qquad \times \binom{n-x-y}{n-m} \binom{-x-\frac{1}{4}}{n-m} \binom{-y-\frac{1}{4}}{n-m} \Big/ \binom{n}{m}^2 \\ &= \binom{2n}{n} \binom{\frac{n-x+y}{2}-\frac{3}{4}}{n} \binom{\frac{n+x-y}{2}-\frac{3}{4}}{n} = \binom{2n}{n}^2 \binom{n+x-y-\frac{1}{2}}{2n} (-\tfrac{1}{4})^n. \end{aligned}$$

For $x=y=0$ we obtain

$$\sum_{m=0}^{n} \binom{-\frac{1}{4}}{m}^2 \binom{-\frac{1}{4}}{n-m}^2 = \binom{2n}{n} \binom{\frac{n}{2}-\frac{3}{4}}{n}^2 = \binom{2n}{n}^2 \binom{n-\frac{1}{2}}{2n} (-\tfrac{1}{4})^n.$$

This solves the problem.

9.9 Peter Larcombe's Problem

In 2005, Peter Larcombe asked whether the following formula is true, and how to prove it:

$$16^n \sum_{k=0}^{2n} 4^k \binom{-\frac{1}{2}}{k} \binom{\frac{1}{2}}{k} \binom{-2k}{2n-k} = (4n+1) \binom{2n}{n}^2. \qquad (9.55)$$

Using the technique of getting a standard form, one can change the left side to

$$\frac{2 \cdot 4^{2n}}{(2n)!^2} \sum_{k=0}^{2n} \binom{2n}{k} [-2n]_k [\tfrac{1}{2}]_k [2n]_{2n-k} [2n-1]_{2n-k} (-1)^k.$$

Using the formula (9.8) on the sum (without the front factor), we get

$$\sum_{k=0}^{2n} \binom{2n}{k} [\tfrac{1}{2}]_k [\tfrac{1}{2}]_k [-\tfrac{1}{2}]_{2n-k} [-1\tfrac{1}{2}]_{2n-k} (-1)^k.$$

This expression is a generalized Dixon formula, so we have the tools to compute it. Let us define

$$S_n(a,b;p,q) = \sum_{k=0}^{n} \binom{n}{k} [a+p]_k [b+q]_k [a]_{n-k} [b]_{n-k} (-1)^k.$$

Then we are after

$$S_{2n}(-\tfrac{1}{2}, -1\tfrac{1}{2}; 1, 2).$$

Now the general recursion formula (9.18) allows us to write this as

$$S_{2n}(-\tfrac{1}{2}, -1\tfrac{1}{2}; 2, 2) + nS_{2n-1}(-\tfrac{1}{2}, -1\tfrac{1}{2}; 1, 1).$$

Both sums are quasi-symmetric Dixon sums to be evaluated by Theorem 9.5 with the formula (9.15) as

$$\begin{aligned} S_{2n}(-\tfrac{1}{2}, -1\tfrac{1}{2}; 2, 2) = {} & [2n]_n [2n-1]_n [-\tfrac{1}{2}]_n [-1\tfrac{1}{2}]_n \\ & + [2n]_{n+1} [2n-1]_{n+1} [-\tfrac{1}{2}]_{n-1} [-1\tfrac{1}{2}]_{n-1}, \end{aligned}$$

and

$$S_{2n-1}(-\tfrac{1}{2}, -1\tfrac{1}{2}; 1, 1) = [2n-1]_n [2n-1]_n [-\tfrac{1}{2}]_{n-1} [-1\tfrac{1}{2}]_{n-1}.$$

Now a tedious computation leads to the result

$$[2n]_n^4 \tfrac{1}{2} (\tfrac{1}{4})^{2n} (4n+1),$$

proving (9.55).

Chapter 10

Sums of Type II(4,4,±1)

We do not know any indefinite formulas of Type II(4,4,±1) , so we rectify the limits and consider a sum of terms

$$S(c,n) = \sum_{k=0}^{n} s(c,n,k), \tag{10.1}$$

with quotient

$$q_t(k) = \frac{(n-k)(\alpha_1 - k)(\alpha_2 - k)(\alpha_3 - k)}{(-1-k)(\beta_1 - k)(\beta_2 - k)(\beta_3 - k)} z.$$

We prefer to define $a_j = \alpha_j$ and $b_j = n-1-\beta_j$, $(j = 1,2,3)$, and hence redefine the sum as

$$S_n(a_1, a_2, a_3, b_1, b_2, b_3) = \sum_{k=0}^{n} \binom{n}{k} [a_1]_k [a_2]_k [a_3]_k [b_1]_{n-k} [b_2]_{n-k} [b_3]_{n-k} z^k.$$

10.1 Sum Formulas for $z = 1$

In the case of $z = 1$, we know two important formulas with three arguments and one excess. The first is the symmetric formula of H. W. Gould [Gould 72b, (23.1)]:

Theorem 10.1. *For any $a, b, c \in \mathbb{C}$ and $n \in \mathbb{N}$ we have that if*

$$a + b + c = n - \tfrac{1}{2},$$

then

$$\sum_{k=0}^{n} \binom{n}{k} [a]_k [b]_k [c]_k [a]_{n-k} [b]_{n-k} [c]_{n-k} = \left(\tfrac{1}{4}\right)^n [2a]_n [2b]_n [2c]_n.$$

This can be generalized to allow excess p:

Theorem 10.2. *For any $a, b, c \in \mathbb{C}$ and $n, p \in \mathbb{N}_0$ we have that if*

$$p = a + b + c - n + \tfrac{1}{2}, \tag{10.2}$$

then

$$\begin{aligned}\sum_{k=0}^{n} \binom{n}{k} [a]_k [b]_k [c]_k [a]_{n-k} [b]_{n-k} [c]_{n-k} &= (-1)^p 2^n [a]_{\lceil \frac{n}{2} \rceil} [b]_{\lceil \frac{n}{2} \rceil} [c]_{\lceil \frac{n}{2} \rceil} \\ &\times \left[a - p - \tfrac{1}{2}\right]_{\lfloor \frac{n}{2} \rfloor - p} \left[b - p - \tfrac{1}{2}\right]_{\lfloor \frac{n}{2} \rfloor - p} \left[c - p - \tfrac{1}{2}\right]_{\lfloor \frac{n}{2} \rfloor - p} \\ &\times \sum_{i=0}^{p} \binom{p}{i} \left[\lfloor \tfrac{n}{2} \rfloor\right]_i [c]_i \left[c - \lceil \tfrac{n}{2} \rceil\right]_i \\ &\times \left[p - c - \tfrac{1}{2}\right]_{p-i} \left[a - \lfloor \tfrac{n}{2} \rfloor - \tfrac{1}{2}\right]_{p-i} \left[b - \lfloor \tfrac{n}{2} \rfloor - \tfrac{1}{2}\right]_{p-i}.\end{aligned}$$

We need a lemma to prove these formulas.

Lemma 10.3. *For any $a, b, x, y \in \mathbb{C}$ and $n \in \mathbb{N}$ we have*

$$\begin{aligned}\sum_{k=0}^{n} \binom{n}{k} [x]_k [y]_{n-k} [a]_k [a]_{n-k} [b]_k [b]_{n-k} &= (-1)^n \sum_{j=0}^{\lfloor \frac{n}{2} \rfloor} \frac{[n]_{2j}}{j!} [x]_j [y]_j \\ &\times [n - 1 - a - b]_j [a]_{n-j} [b]_{n-j} [n - 1 - x - y]_{n-2j} (-1)^j.\end{aligned} \tag{10.3}$$

Proof: We use the Pfaff–Saalschütz identity (9.1) to write

$$[a]_k [b]_k [a]_{n-k} [b]_{n-k} = \sum_{j=0}^{k} \binom{k}{j} [n-k]_j [n - 1 - a - b]_j [a]_{n-j} [b]_{n-j} (-1)^j. \tag{10.4}$$

Substitution of this into the left side of (10.3), after changing the order of summation, yields

$$\begin{aligned}\sum_{j=0}^{n} \binom{n}{j} &[n - 1 - a - b]_j [a]_{n-j} [b]_{n-j} [x]_j [y]_j (-1)^j \\ &\times \sum_{k=j}^{n} \binom{n-j}{k-j} [n-k]_j [x-j]_{k-j} [y-j]_{n-j-k}.\end{aligned} \tag{10.5}$$

After writing $\binom{n-j}{k-j}[n-k]_j = \binom{n-2j}{k-j}[n-j]_j$, we can apply the Chu–Vandermonde formula (8.3) to the inner sum and get

$$[n-j]_j [x + y - 2j]_{n-2j}.$$

Then we write

$$[x + y - 2j]_{n-2j} = (-1)^n [n - 1 - x - y]_{n-2j},$$

and eventually

$$\binom{n}{j} [n - j]_j = \frac{[n]_{2j}}{j!}.$$

Substituting all the changes, the sum becomes the right side of (10.3). □

Proof of Theorem 10.2: We use Lemma 10.3 with $n - 1 - a - b = c - \frac{1}{2} - p$ and $x = y = c$ to write the sum as

$$(-1)^n \sum_{j=0}^{\lfloor \frac{n}{2} \rfloor} \frac{[n]_{2j}}{j!} [c]_j^2 [c - \tfrac{1}{2} - p]_j [a]_{n-j} [b]_{n-j} [n - 1 - 2c]_{n-2j} (-1)^j.$$

Then

$$\begin{aligned}
[n - 1 - 2c]_{n-2j} (-1)^n &= [2c - 2j]_{n-2j} \\
&= 2^{n-2j} [c - j]_{\lceil \frac{n}{2} \rceil - j} \left[c - j - \tfrac{1}{2}\right]_{\lfloor \frac{n}{2} \rfloor - j}, \\
[c]_j [c - j]_{\lceil \frac{n}{2} \rceil - j} &= [c]_{\lceil \frac{n}{2} \rceil}, \\
[c - \tfrac{1}{2} - p]_j \left[c - j - \tfrac{1}{2}\right]_{\lfloor \frac{n}{2} \rfloor - j} &= \left[c - \tfrac{1}{2} - p\right]_{\lfloor \frac{n}{2} \rfloor - p} \left[c - j - \tfrac{1}{2}\right]_p, \\
\left[c - j - \tfrac{1}{2}\right]_p &= (-1)^p \left[p - c - \tfrac{1}{2} + j\right]_p.
\end{aligned}$$

Eventually we write

$$\frac{[n]_{2j}}{j!} = 2^{2j} \binom{\lfloor \frac{n}{2} \rfloor}{j} \left[\lceil \tfrac{n}{2} \rceil - \tfrac{1}{2}\right]_j.$$

Substituting all the changes, the sum becomes

$$\begin{aligned}
&2^n [c]_{\lceil \frac{n}{2} \rceil} \left[c - \tfrac{1}{2} - p\right]_{\lfloor \frac{n}{2} \rfloor - p} (-1)^p \\
&\quad \times \sum_{j=0}^{\lfloor \frac{n}{2} \rfloor} \binom{\lfloor \frac{n}{2} \rfloor}{j} \left[\lceil \tfrac{n}{2} \rceil - \tfrac{1}{2}\right]_j [c]_j [a]_{n-j} [b]_{n-j} (-1)^j \left[p - c - \tfrac{1}{2} + j\right]_p.
\end{aligned}$$

Now we apply the Chu–Vandermonde convolution (8.3) to write the last sum as

$$\left[p - c - \tfrac{1}{2} + j\right]_p = \sum_{h=0}^{p} \binom{p}{h} \left[p - c - \tfrac{1}{2}\right]_{p-h} [j]_h,$$

and use the formula (2.8) to write

$$\binom{\lfloor \frac{n}{2} \rfloor}{j} [j]_h = \left[\lfloor \tfrac{n}{2} \rfloor\right]_h \binom{\lfloor \frac{n}{2} \rfloor - h}{j - h}.$$

Eventually we change the order of summation to get

$$2^n [c]_{\lceil \frac{n}{2} \rceil} \left[c - \tfrac{1}{2} - p\right]_{\lfloor \frac{n}{2} \rfloor - p} (-1)^p \sum_{h=0}^{p} \binom{p}{h} \left[p - c - \tfrac{1}{2}\right]_{p-h} \left[\lfloor \tfrac{n}{2} \rfloor\right]_h$$
$$\times \sum_{j=h}^{\lfloor \frac{n}{2} \rfloor} \binom{\lfloor \frac{n}{2} \rfloor - h}{j - h} \left[\lceil \tfrac{n}{2} \rceil - \tfrac{1}{2}\right]_j [c]_j [a]_{n-j} [b]_{n-j} (-1)^j.$$

The inner sum can be written as

$$\left[\lceil \tfrac{n}{2} \rceil - \tfrac{1}{2}\right]_h [c]_h [a]_{\lceil \frac{n}{2} \rceil} [b]_{\lceil \frac{n}{2} \rceil} (-1)^h \sum_{j=h}^{\lfloor \frac{n}{2} \rfloor} \binom{\lfloor \frac{n}{2} \rfloor - h}{j - h}$$
$$\times \left[\lceil \tfrac{n}{2} \rceil - \tfrac{1}{2} - h\right]_{j-h} [c - h]_{j-h}$$
$$\times \left[a - \lceil \tfrac{n}{2} \rceil\right]_{\lfloor \frac{n}{2} \rfloor - j} \left[b - \lceil \tfrac{n}{2} \rceil\right]_{\lfloor \frac{n}{2} \rfloor - j} (-1)^{j-h}.$$

This sum satisfies the condition for applying the generalized Pfaff–Saalschütz identity (9.3) with excess,

$$\lceil \tfrac{n}{2} \rceil - \tfrac{1}{2} - h + c - h + a - \lceil \tfrac{n}{2} \rceil + b - \lceil \tfrac{n}{2} \rceil - \lfloor \tfrac{n}{2} \rfloor + h + 1 = p - h,$$

so we can write it

$$\sum_{i=0}^{p-h} \binom{p-h}{i} \left[\lfloor \tfrac{n}{2} \rfloor - h\right]_i [c - h]_i \left[\lceil \tfrac{n}{2} \rceil - \tfrac{1}{2} - h + a - \lceil \tfrac{n}{2} \rceil - p + h\right]_{\lfloor \frac{n}{2} \rfloor - h - i}$$
$$\times \left[\lceil \tfrac{n}{2} \rceil - \tfrac{1}{2} - h + b - \lceil \tfrac{n}{2} \rceil - p + h\right]_{\lfloor \frac{n}{2} \rfloor - h - i} (-1)^i$$
$$= \sum_{i=0}^{p-h} \binom{p-h}{i} \left[\lfloor \tfrac{n}{2} \rfloor - h\right]_i [c - h]_i \left[a - p - \tfrac{1}{2}\right]_{\lfloor \frac{n}{2} \rfloor - h - i}$$
$$\times \left[b - p - \tfrac{1}{2}\right]_{\lfloor \frac{n}{2} \rfloor - h - i} (-1)^i.$$

Now we want to rewrite

$$\binom{p}{h}\binom{p-h}{i} = \binom{p}{h+i}\binom{h+i}{h},$$
$$\left[\lfloor \tfrac{n}{2} \rfloor\right]_h \left[\lfloor \tfrac{n}{2} \rfloor - h\right]_i = \left[\lfloor \tfrac{n}{2} \rfloor\right]_{h+i}, \quad [c]_h [c - h]_i = [c]_{h+i}.$$

So the sum becomes

$$\begin{aligned}
&2^n [a]_{\lceil \frac{n}{2} \rceil} [b]_{\lceil \frac{n}{2} \rceil} [c]_{\lceil \frac{n}{2} \rceil} \left[c - \tfrac{1}{2} - p\right]_{\lfloor \frac{n}{2} \rfloor - p} (-1)^p \\
&\quad \times \left[a - p - \tfrac{1}{2}\right]_{\lfloor \frac{n}{2} \rfloor - p} \left[b - p - \tfrac{1}{2}\right]_{\lfloor \frac{n}{2} \rfloor - p} \\
&\quad \times \sum_{h=0}^{p} \left[p - c - \tfrac{1}{2}\right]_{p-h} \left[\lceil \tfrac{n}{2} \rceil - \tfrac{1}{2}\right]_h \\
&\quad \times \sum_{i=0}^{p-h} \binom{p}{h+i} \binom{h+i}{h} \left[\lfloor \tfrac{n}{2} \rfloor\right]_{h+i} [c]_{h+i} \left[a - \lfloor \tfrac{n}{2} \rfloor - \tfrac{1}{2}\right]_{p-h-i} \\
&\quad \times \left[b - \lfloor \tfrac{n}{2} \rfloor - \tfrac{1}{2}\right]_{p-h-i} (-1)^{h+i}.
\end{aligned}$$

Now we only need to change the summation variable. We write i for $i + h$ and obtain

$$\begin{aligned}
&2^n [a]_{\lceil \frac{n}{2} \rceil} [b]_{\lceil \frac{n}{2} \rceil} [c]_{\lceil \frac{n}{2} \rceil} \left[c - \tfrac{1}{2} - p\right]_{\lfloor \frac{n}{2} \rfloor - p} (-1)^p \\
&\quad \times \left[a - p - \tfrac{1}{2}\right]_{\lfloor \frac{n}{2} \rfloor - p} \left[b - p - \tfrac{1}{2}\right]_{\lfloor \frac{n}{2} \rfloor - p} \\
&\quad \times \sum_{h=0}^{p} \left[p - c - \tfrac{1}{2}\right]_{p-h} \left[\lceil \tfrac{n}{2} \rceil - \tfrac{1}{2}\right]_h \\
&\quad \times \sum_{i=h}^{p} \binom{p}{i} \binom{i}{h} \left[\lfloor \tfrac{n}{2} \rfloor\right]_{i} [c]_{i} \left[a - \lfloor \tfrac{n}{2} \rfloor - \tfrac{1}{2}\right]_{p-i} \left[b - \lfloor \tfrac{n}{2} \rfloor - \tfrac{1}{2}\right]_{p-i} (-1)^{i}.
\end{aligned}$$

Now it is time to change the order of summation again to get

$$\begin{aligned}
&2^n [a]_{\lceil \frac{n}{2} \rceil} [b]_{\lceil \frac{n}{2} \rceil} [c]_{\lceil \frac{n}{2} \rceil} \left[c - \tfrac{1}{2} - p\right]_{\lfloor \frac{n}{2} \rfloor - p} (-1)^p \\
&\quad \times \left[a - p - \tfrac{1}{2}\right]_{\lfloor \frac{n}{2} \rfloor - p} \left[b - p - \tfrac{1}{2}\right]_{\lfloor \frac{n}{2} \rfloor - p} \\
&\quad \times \sum_{i=0}^{p} \binom{p}{i} \left[\lfloor \tfrac{n}{2} \rfloor\right]_{i} [c]_{i} \left[a - \lfloor \tfrac{n}{2} \rfloor - \tfrac{1}{2}\right]_{p-i} \left[b - \lfloor \tfrac{n}{2} \rfloor - \tfrac{1}{2}\right]_{p-i} (-1)^{i} \\
&\quad \times \sum_{h=0}^{i} \binom{i}{h} \left[p - c - \tfrac{1}{2}\right]_{p-h} \left[\lceil \tfrac{n}{2} \rceil - \tfrac{1}{2}\right]_h .
\end{aligned}$$

The inner sum is subject to summation using the Chu–Vandermonde formula (8.3); it becomes

$$\begin{aligned}
&\sum_{h=0}^{i} \binom{i}{h} \left[p - c - \tfrac{1}{2}\right]_{p-h} \left[\lceil \tfrac{n}{2} \rceil - \tfrac{1}{2}\right]_h \\
&\qquad = \left[p - c - \tfrac{1}{2}\right]_{p-i} \sum_{h=0}^{i} \binom{i}{h} \left[i - c - \tfrac{1}{2}\right]_{i-h} \left[\lceil \tfrac{n}{2} \rceil - \tfrac{1}{2}\right]_h
\end{aligned}$$

$$= \left[p - c - \tfrac{1}{2}\right]_{p-i} \left[\lfloor \tfrac{n}{2} \rfloor + i - c - 1\right]_i$$
$$= \left[p - c - \tfrac{1}{2}\right]_{p-i} \left[c - \lfloor \tfrac{n}{2} \rfloor\right]_i (-1)^i.$$

This proves the formula in the theorem. □

For $p = 0$ we get Theorem 10.1.

Theorem 10.4. *For any $a, b, x, y \in \mathbb{C}$ and $n, p \in \mathbb{N}_0$, we have*

$$\sum_{k=0}^{n} \binom{n}{k} [x]_k [y]_{n-k} [a]_k [a]_{n-k} [b]_k [b]_{n-k}$$
$$= \sum_{k=0}^{n} \binom{n}{k} [x]_k [x + p - k]_{n-k} [a]_k [a]_{n-k} [b]_k [b]_{n-k} (-1)^{n-k}$$
$$= (-1)^n \sum_{j=\lceil \frac{n-p}{2} \rceil}^{\lfloor \frac{n}{2} \rfloor} \frac{[n]_{2j}}{j!} [x]_j [y]_j [n - 1 - a - b]_j [a]_{n-j} [b]_{n-j} [p]_{n-2j} (-1)^j,$$

where $x + y = n - 1 - p$.

Proof: This follows immediately from Lemma 10.3. □

The special case of $x = y = \frac{n}{2} - 1$, which gives $p = 1$, was found by L. J. Slater [Slater 66, (III.22)].

There is a balanced identity of type II$(4, 4, 1)$.

Theorem 10.5. *For any $a, b, c \in \mathbb{C}$ and $n \in \mathbb{N}$, satisfying the condition $b + c = -\frac{1}{2}$, we have*

$$\sum_{k=0}^{n} \binom{n}{k} [n + 2a]_k [n - 2a]_{n-k} [b + a]_k [b - a]_{n-k} [c + a]_k [c - a]_{n-k}$$
$$= \left(-\tfrac{1}{4}\right)^n [2n]_n [2b + n]_n.$$

Proof: This is a corollary to Theorem 11.4 with $p = 0$ and $d = \frac{n-1}{2}$. □

We also have a well-balanced identity of type II$(4, 4, 1)$, by L. J. Slater [Slater 66, (III.26)].

Theorem 10.6. *For any $a, b \in \mathbb{C}$ and $n \in \mathbb{N}$ we have*

$$\sum_{k=0}^{n} \binom{n}{k} [n + 2a]_k [n - 2a]_{n-k} [b + a]_k [b - a]_{n-k} (n + 2a - 2k)(-1)^k$$
$$= 2^{2n+1} \left[\tfrac{n}{2} + a\right]_{n+1} \left[\tfrac{n-1}{2} + b\right]_n (-1)^n.$$

Proof: This is a corollary to Theorem 10.1, obtained by inserting $c = \frac{n-1}{2}$ into (10.1). □

10.2 Sum Formulas for $z = -1$

We also have a well-balanced identity of type II$(4, 4, -1)$, by L. J. Slater [Slater 66, (III.11)].

Theorem 10.7. *For any $a, b \in \mathbb{C}$ and $n \in \mathbb{N}$ we have*

$$\sum_{k=0}^{n} \binom{n}{k} [n+2a]_k [n-2a]_{n-k} [b+a]_k [b-a]_{n-k} (n+2a-2k) = [n+2a]_{2n+1}.$$

Proof: This theorem follows as a corollary to a well-balanced formula of type II$(5, 5, 1)$ in Chapter 11. Consider the equation (11.3) as a polynomial identity in the variable, c, and compare the coefficients of the term c^n. □

There are a couple of general formulas, neither symmetric nor balanced. One family of these is the following.

Theorem 10.8. *For any $a, b \in \mathbb{C}$ and any $n, p \in \mathbb{N}_0$ we have*

$$\begin{aligned}\sum_{k=0}^{n} \binom{n}{k} &\left[\tfrac{a}{2}\right]_k \left[\tfrac{a-1}{2}\right]_k [n+p-b]_k \left[\tfrac{b}{2}\right]_{n-k} \left[\tfrac{b-1}{2}\right]_{n-k} [n-1-a]_{n-k} \\ &= \left(\tfrac{1}{4}\right)^n [a]_n [b]_{n+p+1} [2n-a-b-1]_{n-p} \\ &\qquad \times \sum_{j=0}^{p} \binom{p}{j} [n-a-b-1-p]_{p-j} [n]_j [b-p-1-j]_{-p-1} (-1)^j.\end{aligned}$$

Proof: We write the product

$$\begin{aligned}\left[\tfrac{a}{2}\right]_k \left[\tfrac{a-1}{2}\right]_k [n-1-a]_{n-k} &= 4^{-k} [a]_{2k} (-1)^{n-k} [a-k]_{n-k} \\ &= (-1)^n (-4)^{-k} [a]_n [a-k]_k\end{aligned}$$

and the product

$$\begin{aligned}&\left[\tfrac{b}{2}\right]_{n-k} \left[\tfrac{b-1}{2}\right]_{n-k} [n+p-b]_k \\ &\qquad = 4^{k-n} [b]_{2n-2k} (-1)^k [b-p-n-1+k]_k \\ &\qquad = 4^{-n} (-4)^k [b]_{n+p+1} [b-p-n-1+k]_{n-p-1-k}.\end{aligned}$$

The sum then becomes

$$\left(-\tfrac{1}{4}\right)^n [a]_n [b]_{n+p+1} \sum_{k=0}^{n} \binom{n}{k} [a-k]_k [b-p-n-1+k]_{n-p-1-k}.$$

Now we apply the Chu–Vandermonde formula (8.2) to write

$$[a-k]_k = (-1)^k[-a-1+2k]_k = (-1)^k \sum_{j=0}^{k} \binom{k}{j} [2n-a-b-1]_j [b-2n+2k]_{k-j}.$$

Substituting (2.9) into the sum and interchanging the order of summation yields

$$\left(-\tfrac{1}{4}\right)^n [a]_n [b]_{n+p+1} \sum_{j=0}^{n} \binom{n}{j} [2n-a-b-1]_j (-1)^j$$
$$\times \sum_{k=j}^{n} \binom{n-j}{k-j} (-1)^{k-j} [b-p-n-1+k]_{n-p-1-j}.$$

By the Chu–Vandermonde formula (8.8) the inner sum is equal to

$$[b-p-n-1+j]_{-p-1}[n-p-j-1]_{n-j}(-1)^{n-j} = [b-p-n-1+j]_{-p-1}[p]_{n-j},$$

so the sum becomes

$$\left(-\tfrac{1}{4}\right)^n [a]_n [b]_{n+p+1} \sum_{j=0}^{n} \binom{n}{j} [2n-a-b-1]_j [b-p-n-1+j]_{-p-1} [p]_{n-j} (-1)^j.$$

Reversing the direction of summation, splitting the first factor and using $\binom{n}{j}[p]_j = \binom{p}{j}[n]_j$ proves the theorem. □

The theorem for $p = 0$ reduces to the following formula from J. L. Slater [Slater 66, (III.20)]:

Theorem 10.9. *For any $a, b \in \mathbb{C}$ and any $n \in \mathbb{N}$ we have*

$$\sum_{k=0}^{n} \binom{n}{k} \left[\tfrac{a}{2}\right]_k \left[\tfrac{a-1}{2}\right]_k [n-b]_k \left[\tfrac{b}{2}\right]_{n-k} \left[\tfrac{b-1}{2}\right]_{n-k} [n-1-a]_{n-k}$$
$$= \left(\tfrac{1}{4}\right)^n [a]_n [b-1]_n [2n-a-b-1]_n.$$

Furthermore, there is a beautiful formula due to H. L. Krall [Krall 38], with equivalent forms by H. W. Gould [Gould 72b, (11.5) and (11.6)].

In Krall's form it is:

Theorem 10.10. *For all $x \neq y \in \mathbb{C}$ we have*

$$\sum_{k} \binom{n-k}{k} [x]_k [y]_k [n-x-y]_{n-2k} (-1)^k$$
$$= (-1)^n \frac{[x]_{n+1} - [y]_{n+1}}{x-y} = \sum_{k=0}^{n} [x]_k [n-y]_{n-k} (-1)^{n-k}. \qquad (10.6)$$

In our standard form, it is:

Theorem 10.11. *For all $x \neq y \in \mathbb{C}$ we have*

$$\sum_{k=0}^{\lfloor \frac{n}{2} \rfloor} \binom{\lfloor \frac{n}{2} \rfloor}{k} \left[\lceil \tfrac{n}{2} \rceil - \tfrac{1}{2}\right]_k [x]_k [y]_k \left[-\lceil \tfrac{n}{2} \rceil - 1\right]_{\lfloor \frac{n}{2} \rfloor - k}$$
$$\times \left[\tfrac{n-x-y}{2}\right]_{\lceil \frac{n}{2} \rceil - k} \left[\tfrac{n-x-y-1}{2}\right]_{\lfloor \frac{n}{2} \rfloor - k}$$
$$= (-1)^{\lceil \frac{n}{2} \rceil} \left(\tfrac{1}{2}\right)^n \frac{[x]_{n+1} - [y]_{n+1}}{x - y}.$$

Proof: We prove the form (10.6). Let us define

$$S_n(x,y) = \sum_k \binom{n-k}{k} [x]_k [y]_k [n-x-y]_{n-2k} (-1)^k.$$

Then formula (10.6) is obvious for $n = 0, 1$, so we consider $S_n(x,y)$. Splitting the binomial coefficient we obtain

$$S_n(x,y) = \sum_k \binom{n-1-k}{k} [x]_k [y]_k [n-x-y]_{n-2k} (-1)^k$$
$$+ \sum_k \binom{n-1-k}{k-1} [x]_k [y]_k [n-x-y]_{n-2k} (-1)^k$$
$$= (n-x-y) S_{n-1}(x,y) - \sum_k \binom{n-2-k}{k} [x]_{k+1} [y]_{k+1}$$
$$\times [n-2-(x-1)-(y-1)]_{n-2-2k} (-1)^k$$
$$= (n-x-y) S_{n-1}(x,y) - xy S_{n-2}(x-1, y-1).$$

Substitution of the right side of (10.6) gives

$$\frac{(n-x-y)\left([x]_n - [y]_n\right) + xy\left([x-1]_{n-1} - [y-1]_{n-1}\right)}{(-1)^{n-1}(x-y)}$$
$$= \frac{(n-x)x[x-1]_{n-1} - (n-y)y[y-1]_{n-1}}{(-1)^{n-1}(x-y)} = \frac{[x]_{n+1} - [y]_{n+1}}{(-1)^n (x-y)},$$

which is the correct right side of (10.6) for n. □

We can split the formula in Theorem 10.11 into the even and odd cases:

Theorem 10.12. *For all $x \neq y \in \mathbb{C}$ we have*

$$\sum_{k=0}^{n} \binom{n}{k} \left[n - \tfrac{1}{2}\right]_k [x]_k [y]_k \left[-n-1\right]_{n-k} \left[n - \tfrac{x+y}{2}\right]_{n-k} \left[n - \tfrac{x+y+1}{2}\right]_{n-k}$$
$$= \left(-\tfrac{1}{4}\right)^n \frac{[x]_{2n+1} - [y]_{2n+1}}{x-y},$$
$$\sum_{k=0}^{n} \binom{n}{k} \left[n + \tfrac{1}{2}\right]_k [x]_k [y]_k \left[-n-2\right]_{n-k} \left[n - \tfrac{x+y}{2}\right]_{n-k} \left[n - \tfrac{x+y+1}{2}\right]_{n-k}$$
$$= \left(-\tfrac{1}{4}\right)^n \frac{[x]_{2n+2} - [y]_{2n+2}}{(x+y-2n-1)(x-y)}.$$

Chapter 11

Sums of Type II(5,5,1)

11.1 Indefinite Sums

There is a general summation formula of type II$(5, 5, 1)$, namely:

Theorem 11.1. *For any* $a, b, c \in \mathbb{C}$ *we have*

$$\Delta \frac{[a+b+c]_k [a-1]_{k-1} [b-1]_{k-1} [c-1]_{k-1} (-1)^k}{k! [b+c-1]_{k-1} [a+c-1]_{k-1} [a+b-1]_{k-1}}$$
$$= \frac{[a+b+c]_k [a]_k [b]_k [c]_k (a+b+c-2k)(-1)^k}{k! [b+c-1]_k [a+c-1]_k [a+b-1]_k}. \tag{11.1}$$

Proof: Straightforward computation. □

11.2 Symmetric and Balanced Sums

There are quite a few symmetric and balanced formulas of type II$(5, 5, 1)$, and many are well-balanced. We start with a symmetric formula.

Theorem 11.2. *For any* $n, p \in \mathbb{N}_0$ *and* $a, b, c, d \in \mathbb{C}$ *satisfying the condition*

$$a + b + c + d - 3n + 1 = p,$$

we have

$$\sum_{k=0}^{2n} \binom{2n}{k} [a]_k [b]_k [c]_k [d]_k [a]_{2n-k} [b]_{2n-k} [c]_{2n-k} [d]_{2n-k} (-1)^k$$
$$= [2n]_n [a]_n [b]_n [c]_n [d]_n [a+b-n-p]_{n-p} [a+c-n-p]_{n-p}$$
$$\times [b+c-n-p]_{n-p} \sum_{j=0}^{p} \binom{p}{j} [n]_j [d]_j [2n-a-b-c+p-1]_j$$
$$\times [a+b-2n]_{p-j} [a+c-2n]_{p-j} [b+c-2n]_{p-j}$$

$$= [2n]_n[a]_n[b]_n[c]_n[d]_n \sum_{j=0}^{p} \binom{p}{j} [n]_j[d]_j[2n-a-b-c+p-1]_j$$
$$\times [a+b-n-p]_{n-j}[a+c-n-p]_{n-j}[b+c-n-p]_{n-j}. \quad (11.2)$$

Proof: We apply the Pfaff–Saalschütz identity (9.1) to write

$$[a]_k[b]_k[a]_{2n-k}[b]_{2n-k} = \sum_{j=0}^{k} \binom{k}{j} [2n-k]_j[2n-a-b-1]_j[a]_{2n-j}[b]_{2n-j}(-1)^j.$$

Substitution of this into the sum of (11.2) yields, after changing the order of summation,

$$\sum_{j=0}^{2n} \binom{2n}{j} [2n-a-b-1]_j[a]_{2n-j}[b]_{2n-j}\left([c]_j[d]_j\right)^2$$
$$\times \sum_{k=j}^{2n} \binom{2n-j}{k-j} [2n-k]_j[c-j]_{k-j}[d-j]_{k-j}[c-j]_{2n-j-k}$$
$$\times [d-j]_{2n-j-k}(-1)^{k-j}.$$

Now, we have $[2n-k]_j = 0$ for $j > 2n-k$, so we can end summation for $k = 2n-j$. The inner sum then starts

$$\sum_{k=j}^{2n-j} \binom{2n-2j}{2n-k-j} [2n-j]_j \cdots .$$

The inner sum can be written as

$$[2n-j]_j \sum_{k=j}^{2n-j} \binom{2n-2j}{k-j} [c-j]_{k-j}[d-j]_{k-j}[c-j]_{2n-j-k}[d-j]_{2n-j-k}(-1)^{k-j}.$$

We can apply the Dixon identity (9.2) on this sum and write it as

$$[2n-j]_j[c-j]_{n-j}[d-j]_{n-j}[2n-2j-c+j-d+j-1]_{n-j}[2n-2j]_{n-j}.$$

Therefore, using $\binom{2n}{j}[2n-j]_n = [2n]_n\binom{n}{j}$, we get the sum

$$\sum_{j=0}^{2n} \binom{2n}{j} [2n-a-b-1]_j[a]_{2n-j}[b]_{2n-j}[c]_j[d]_j[2n-j]_n[c]_n[d]_n$$
$$\times [2n-c-d-1]_{n-j}$$
$$= [2n]_n[a]_n[b]_n[c]_n[d]_n \sum_{j=0}^{2n} \binom{n}{j} [2n-a-b-1]_j[a-n]_{n-j}$$
$$\times [b-n]_{n-j}[c]_j[d]_j[a+b-n-p]_{n-j}.$$

Now we can write

$$\begin{aligned}
&[a+b-n-p]_{n-j}[2n-a-b-1]_j \\
&\quad = [a+b-n-p]_{n-p}[a+b-2n]_{p-j}[a+b-2n+j]_j(-1)^j \\
&\quad = [a+b-n-p]_{n-p}[a+b-2n+j]_p(-1)^j.
\end{aligned}$$

And we can apply the Chu–Vandermonde Theorem 8.1 to write

$$[a+b-2n+j]_p = \sum_{h=0}^{p} \binom{p}{h} [a+b-2n]_{p-h}[j]_h.$$

Interchanging the order of summation yields

$$\begin{aligned}
&[2n]_n[a]_n[b]_n[c]_n[d]_n[a+b-n-p]_{n-p} \sum_{h=0}^{p} \binom{p}{h} [a+b-2n]_{p-h}(-1)^h \\
&\quad \times [c]_h[d]_h \sum_{j=0}^{2n} \binom{n}{j} [j]_h[a-n]_{n-j}[b-n]_{n-j}[c-h]_{j-h}[d-h]_{j-h}(-1)^{j-h}.
\end{aligned}$$

As we remark that $\binom{n}{j}[j]_h = [n]_h\binom{n-h}{j-h}$ and we can see that the condition for the generalized Pfaff–Saalschütz identity (9.10) yields $c-h+d-h+a-n+b-n-n+h+1 = p-h$, we can write the sum as

$$\begin{aligned}
&[2n]_n[a]_n[b]_n[c]_n[d]_n[a+b-n-p]_{n-p} \sum_{h=0}^{p} \binom{p}{h} [a+b-2n]_{p-h}(-1)^h \\
&\quad \times [c]_h[d]_h[n]_h \sum_{i=0}^{p-h} \binom{p-h}{i} [n-h]_i[d-h]_i \\
&\quad \times [c-h+a-n-(p-h)]_{n-h-i}[c-h+b-n-(p-h)]_{n-h-i}(-1)^i.
\end{aligned}$$

As we can write $\binom{p}{h}\binom{p-h}{i} = \binom{p}{h+i}\binom{h+i}{h}$ and join $[n]_h[n-h]_i = [n]_{h+i}$ and $[d]_h[d-h]_i = [d]_{h+i}$, we can change the variable i to $i-h$ and write the sum as

$$\begin{aligned}
&[2n]_n[a]_n[b]_n[c]_n[d]_n[a+b-n-p]_{n-p} \sum_{h=0}^{p} [a+b-2n]_{p-h}[c]_h \\
&\quad \times \sum_{i=h}^{p} \binom{p}{i}\binom{i}{h} [n]_i[d]_i[c+a-n-p]_{n-i}[c+b-n-p]_{n-i}(-1)^i.
\end{aligned}$$

Interchanging the order of summation yields

$$[2n]_n[a]_n[b]_n[c]_n[d]_n[a+b-n-p]_{n-p}$$
$$\times \sum_{i=0}^{p} \binom{p}{i} [a+b-2n]_{p-i}[n]_i[d]_i[c+a-n-p]_{n-i}[c+b-n-p]_{n-i}(-1)^i$$
$$\times \sum_{h=0}^{i} \binom{i}{h} [a+b-2n-p+i]_{i-h}[c]_h.$$

Here the inner sum is subject to the Chu–Vandermonde convolution (8.3), so we have

$$[2n]_n[a]_n[b]_n[c]_n[d]_n[a+b-n-p]_{n-p} \sum_{i=0}^{p} \binom{p}{i} [a+b-2n]_{p-i}[n]_i[d]_i$$
$$\times [c+a-n-p]_{n-i}[c+b-n-p]_{n-i}(-1)^i[a+b-2n-p+i+c]_i.$$

Rewriting $(-1)^i[a+b-2n-p+i+c]_i = [2n+p-1-a-b-c]_i$ yields the formula (11.2). □

The special case of $p=0$ is due to W. A. Al-Salam [Al-Salam 57], cf. H. W. Gould [Gould 72b, (22.1)]. We proceed with the following well-balanced formula of type II(5, 5, 1), cf. L. J. Slater [Slater 66, (III.13)], although it also follows as a corollary to the formula of J. Dougall in the next chapter, Theorem 12.4. We state the formula in the canonical form (4.24).

Theorem 11.3. *For any $a, b, c \in \mathbb{C}$ and $n \in \mathbb{N}$ we have*

$$\sum_{k=0}^{n} \binom{n}{k} [n+2a]_k[n-2a]_{n-k}[b+a]_k[b-a]_{n-k}[c+a]_k[c-a]_{n-k}$$
$$\times (n+2a-2k)$$
$$= [b+c]_n[n+2a]_{2n+1}. \tag{11.3}$$

Proof: We use the Pfaff–Saalschütz formula (9.1) to write

$$[b+a]_k[c+a]_k = \sum_{j=0}^{k} \binom{k}{j} (-1)^j [b-a-n+k]_{k-j}[c-a-n+k]_{k-j}$$
$$\times [2a+n-k]_j[n-b-c-1]_j. \tag{11.4}$$

Then we combine the factorials,

$$[b-a]_{n-k}[b-a-n+k]_{k-j} = [b-a]_{n-j}, \tag{11.5}$$
$$[c-a]_{n-k}[c-a-n+k]_{k-j} = [c-a]_{n-j}, \tag{11.6}$$
$$[n+2a]_k[n+2a-k]_j = [n+2a]_{k+j}. \tag{11.7}$$

Substitution of (11.4) into (11.3) yields, after interchanging the order of summation using (2.9),

$$\sum_{j=0}^{n} \binom{n}{j} (-1)^j [n-b-c-1]_j [b-a]_{n-j} [c-a]_{n-j}$$
$$\times \sum_{k=j}^{n} \binom{n-j}{k-j} [n+2a]_{k+j} [n-2a]_{n-k} (n+2a-2k).$$

Now we can write

$$[n+2a]_{k+j} = [n+2a]_{2j} [n+2a-2j]_{k-j}$$

and split the factor as

$$n+2a-2k = (n+2a-2j) - 2(k-j).$$

Then we get the inner sum split in two,

$$\begin{aligned}
&\sum_{k=j}^{n} \binom{n-j}{k-j} [n+2a]_{k+j} [n-2a]_{n-k} (n+2a-2k) \\
&\quad = [n+2a]_{2j} (n+2a-2j) \sum_{k=j}^{n} \binom{n-j}{k-j} [n+2a-2j]_{k-j} [n-2a]_{n-k} \\
&\qquad - 2[n+2a]_{2j} (n-j) \sum_{k=j+1}^{n} \binom{n-j-1}{k-j-1} [n+2a-2j]_{k-j} [n-2a]_{n-k} \\
&\quad = [n+2a]_{2j} (n+2a-2j) [2n-2j]_{n-j} \\
&\qquad - 2[n+2a]_{2j} (n-j)(n+2a-2j) [2n-2j-1]_{n-j-1},
\end{aligned}$$

where we have used Chu–Vandermonde (8.2) on the two sums. Now, this vanishes except for $j = n$, in which case the whole sum becomes

$$(-1)^n [n-b-c-1]_n [n+2a]_{2n} (2a-n) = [b+c]_n [n+2a]_{2n+1},$$

which proves the statement. □

There exists a family of balanced formulas, too:

Theorem 11.4. *For any $n \in \mathbb{N}$ and $a, b, c, d \in \mathbb{C}$ satisfying the condition*

$$p = b + c + d - \tfrac{n}{2} + 1 \in \mathbb{N}_0,$$

we have

$$\begin{aligned}
&\sum_{k=0}^{n}\binom{n}{k}[n+2a]_k[b+a]_k[c+a]_k[d+a]_k \\
&\qquad\times[n-2a]_{n-k}[b-a]_{n-k}[c-a]_{n-k}[d-a]_{n-k}(-1)^k \\
&=4^n\left[\tfrac{n-1}{2}+a\right]_n\left[b+\tfrac{n}{2}-p\right]_{n-p}[b+d-p]_{n-p}\left[d+\tfrac{n}{2}-p\right]_{n-p} \\
&\qquad\times\sum_{j=0}^{p}\binom{p}{j}[n]_j[c+a]_j[c-a]_j[b+d-n]_{p-j}\left[b-\tfrac{n}{2}\right]_{p-j}\left[d-\tfrac{n}{2}\right]_{p-j} \\
&=4^n\left[\tfrac{n-1}{2}+a\right]_n\sum_{j=0}^{p}\binom{p}{j}[n]_j[c+a]_j[c-a]_j[b+d-p]_{n-j}\left[b+\tfrac{n}{2}-p\right]_{n-j} \\
&\qquad\times\left[d+\tfrac{n}{2}-p\right]_{n-j}.
\end{aligned} \tag{11.8}$$

Proof: This theorem is a corollary to the generalization of Dougall's theorem, Theorem 12.6. We choose $e=\frac{n}{2}$ in (12.14) and remark that

$$\left[\tfrac{n}{2}+a\right]_k\left[\tfrac{n}{2}-a\right]_{n-k}(n+2a-2k)=(-1)^{n-k}2\left[\tfrac{n}{2}+a\right]_{n+1}.$$

Dividing (12.5) by this constant yields (11.8). □

The special case of (11.8) for $p=0$ was found by S. Ramanujan, cf. [Hardy 23, p. 493].

Chapter 12

Other Type II Sums

12.1 Type II(6,6,$\pm$1)

There is a single well-balanced formula of type II(6,6,$\pm$1) and three others. But we have a kind of general transformation available from type $\mathrm{II}(6,6,-1)$ to type $\mathrm{II}(3,3,1)$.

Theorem 12.1. *For any $a, b, c, d \in \mathbb{C}$ and $n \in \mathbb{N}$ we have*

$$\begin{aligned}
&\sum_{k=0}^{n} \binom{n}{k} [n+2a]_k [b+a]_k [c+a]_k [d+a]_k \\
&\qquad \times [n-2a]_{n-k} [b-a]_{n-k} [c-a]_{n-k} [d-a]_{n-k} (n+2a-2k) \\
&= [n+2a]_{2n+1} \sum_{j=0}^{n} \binom{n}{j} [b+a]_j [n-c-d-1]_j [c-a]_{n-j} [d-a]_{n-j} (-1)^j.
\end{aligned} \tag{12.1}$$

Proof: We apply the Pfaff–Saalschütz formula (9.1) to write

$$\begin{aligned}
[c+a]_k [d+a]_k = &\sum_{j=0}^{k} \binom{k}{j} [c-a-n+k]_{k-j} [d-a-n+k]_{k-j} \\
&\times [n+2a-k]_j [n-c-d-1]_j (-1)^j.
\end{aligned}$$

Then we combine the factorials,

$$\begin{aligned}
[c-a]_{n-k} [c-a-n+k]_{k-j} &= [c-a]_{n-j}, \\
[d-a]_{n-k} [d-a-n+k]_{k-j} &= [d-a]_{n-j}, \\
[n+2a]_k [n+2a-k]_j &= [n+2a]_{k+j} = [n+2a]_{2j} [n+2a-2j]_{k-j},
\end{aligned}$$

to write the left sum of (12.1), after changing the order of summation, as

$$\sum_{j=0}^{n}\binom{n}{j}[n+2a]_{2j}[c-a]_{n-j}[d-a]_{n-j}[n-c-d-1]_j[b+a]_j(-1)^j$$
$$\times\sum_{k=j}^{n}\binom{n-j}{k-j}[n+2a-2j]_{k-j}[b+a-j]_{k-j}$$
$$\times[n-2a]_{n-k}[b-a]_{n-k}(n+2a-2k). \tag{12.2}$$

The inner sum has the form of (10.2), if we replace n with $n-j$, a with $a-\frac{j}{2}$ and b with $b-\frac{j}{2}$. Then Theorem 10.4 gives us

$$\sum_{k=j}^{n}\binom{n-j}{k-j}[n+2a-2j]_{k-j}[b+a-j]_{k-j}$$
$$\times[n-2a]_{n-k}[b-a]_{n-k}(n+2a-2k)$$
$$=[n-j+2a-j]_{2n-2j+1},$$

which, inserted in (12.2), gives

$$[n+2a]_{2n+1}\sum_{j=0}^{n}\binom{n}{j}[b+a]_j[n-c-d-1]_j[c-a]_{n-j}[d-a]_{n-j}(-1)^j.$$

□

This transformation allows no nice formulas of type II$(6,6,-1)$.

The well-balanced formula of type II$(6,6,1)$ is:

Theorem 12.2. *For any $n\in\mathbb{N}$ and $a,b,c,d\in\mathbb{C}$ satisfying the condition $b+c+d=\frac{n-1}{2}$, we have*

$$\sum_{k=0}^{n}\binom{n}{k}(-1)^k[n+2a]_k[b+a]_k[c+a]_k[d+a]_k$$
$$\times[n-2a]_{n-k}[b-a]_{n-k}[c-a]_{n-k}[d-a]_{n-k}(n+2a-2k)$$
$$=(-1)^n2^{2n+1}[a+\tfrac{n}{2}]_{n+1}[b+\tfrac{n-1}{2}]_n[c+\tfrac{n-1}{2}]_n[d+\tfrac{n-1}{2}]_n. \tag{12.3}$$

Proof: This follows as a corollary to Theorem 12.4 by choosing $e=\frac{n-1}{2}$.

□

I. M. Gessel proved three formulas of this type [Gessel 95a, 19.1a, 19.2a, 19.3a]:

$$\begin{aligned}\sum_{k=0}^{n}\binom{n}{k}&[-b-1]_k\left[-a-b-\tfrac{1}{2}\right]_k[a+2b-n-1]_k[b]_k\left[n+\tfrac{1}{2}\right]_{n-k}\\&\times[n-2a-3b]_{n-k}[3n+1-b]_{n-k}\\&\times[2a+3b-n-1]_{n-k}(b-1-3k)(-1)^k\\&=[a+b-1]_n[b-1-n]_{2n+1}\left[a+2b-\tfrac{1}{2}\right]_n(-1)^n,\end{aligned}\tag{12.4}$$

$$\begin{aligned}\sum_{k=0}^{n}\binom{n}{k}&[1-b-a]_k[1-a-2b-n]_k\left[b-\tfrac{1}{2}\right]_k[b+a-1]_k\left[n-\tfrac{1}{2}\right]_{n-k}\\&\times[3n-1+a+b]_{n-k}[n-2+3b+a]_{n-k}[1-a-3b-n]_{n-k}\\&\times(a+b-1+3k)(-1)^k\\&=(a+b-1)\left[\tfrac{1}{2}-a-2b\right]_n[-b]_n[-a-b-n]_{2n}(-1)^n,\end{aligned}\tag{12.5}$$

$$\begin{aligned}\sum_{k=0}^{n}\binom{n}{k}&[-b]_k[1-2a]_k[-b-a-n]_k\left[\tfrac{1}{2}-b-a\right]_k\\&\times[-n-1]_{n-k}[2(b+a)+3n-1]_{n-k}\left[n-1+\tfrac{b}{2}+a\right]_{n-k}\\&\times\left[n-\tfrac{1}{2}+\tfrac{b}{2}+a\right]_{n-k}(2(b+a)-1+3k)(-1)^k\\&=-2[-a]_n[2n+b]_n[-2(b+a+n)]_n\left[\tfrac{1}{2}-b-a\right]_{n+1}.\end{aligned}\tag{12.6}$$

12.2 Type II(7,7,1)

There are only a few well-balanced formulas of type II$(7,7,1)$, and a couple of others. But we have the following general transformation available, which we will use to prove J. Dougall's formula.

Theorem 12.3. *For any $a, b, c, d, e \in \mathbb{C}$ and $n \in \mathbb{N}$ we have*

$$\begin{aligned}\sum_{k=0}^{n}\binom{n}{k}&[n+2a]_k[b+a]_k[c+a]_k[d+a]_k[e+a]_k\\&\times[n-2a]_{n-k}[b-a]_{n-k}[c-a]_{n-k}[d-a]_{n-k}[e-a]_{n-k}(n+2a-2k)\\&=[n+2a]_{2n+1}(-1)^n\sum_{j=0}^{n}\binom{n}{j}[b+a]_j[c+a]_j[n-d-e-1]_j\\&\times[d-a]_{n-j}[e-a]_{n-j}[n-b-c-1]_{n-j}.\end{aligned}\tag{12.7}$$

Proof: We apply the Pfaff–Saalschütz formula (9.1) to write

$$[d+a]_k[e+a]_k = \sum_{j=0}^{k}\binom{k}{j}[d-a-n+k]_{k-j}[e-a-n+k]_{k-j} \\ \times [n+2a-k]_j[n-d-e-1]_j(-1)^j. \quad (12.8)$$

Then we combine the factorials,

$$[d-a]_{n-k}[d-a-n+k]_{k-j} = [d-a]_{n-j},$$
$$[e-a]_{n-k}[e-a-n+k]_{k-j} = [e-a]_{n-j},$$
$$[n+2a]_k[n+2a-k]_j = [n+2a]_{k+j} = [n+2a]_{2j}[n+2a-2j]_{k-j},$$

to write the left sum of (12.7), after changing the order of summation, as

$$\sum_{j=0}^{n}\binom{n}{j}[n+2a]_{2j}[d-a]_{n-j}[e-a]_{n-j}[n-d-e-1]_j[b+a]_j \\ \times [c+a]_j(-1)^j\sum_{k=j}^{n}\binom{n-j}{k-j}[n+2a-2j]_{k-j}[b+a-j]_{k-j} \\ \times [c+a-j]_{k-j}[n-2a]_{n-k}[b-a]_{n-k}[c-a]_{n-k}(n+2a-2k). \quad (12.9)$$

The inner sum has the form of (11.2), if we replace n with $n-j$, a with $a-\frac{j}{2}$, b with $b-\frac{j}{2}$ and c with $c-\frac{c}{2}$. Then Theorem 11.2 gives us

$$\sum_{k=j}^{n}\binom{n-j}{k-j}[n+2a-2j]_{k-j}[b+a-j]_{k-j}[c+a-j]_{k-j} \\ \times [n-2a]_{n-k}[b-a]_{n-k}[c-a]_{n-k}(n+2a-2k) \\ = [n-j+2a-j]_{2n-2j+1}[b+c-j]_{n-j},$$

which, inserted into (12.9), gives

$$[n+2a]_{2n+1}(-1)^n\sum_{j=0}^{n}\binom{n}{j}[b+a]_j[c+a]_j[n-d-e-1]_j[d-a]_{n-j} \\ \times [e-a]_{n-j}[n-b-c-1]_{n-j}. \quad (12.10)$$

□

The following well-known formula was proved in 1907 by J. Dougall [Dougall 07], see also [Gould 72b, (71.1) and (97.1)].

Theorem 12.4 (Dougall's Formula). *For any $n \in \mathbb{N}$ and $a, b, c, d, e \in \mathbb{C}$ satisfying the condition*

$$b + c + d + e = n - 1, \tag{12.11}$$

we have

$$\begin{aligned}
&\sum_{k=0}^{n} \binom{n}{k} [n+2a]_k [b+a]_k [c+a]_k [d+a]_k [e+a]_k [n-2a]_{n-k} \\
&\qquad \times [b-a]_{n-k} [c-a]_{n-k} [d-a]_{n-k} [e-a]_{n-k} (n+2a-2k) \\
&\quad = [n+2a]_{2n+1} [b+e]_n [c+e]_n [d+e]_n \\
&\quad = (-1)^n [n+2a]_{2n+1} [b+c]_n [c+d]_n [d+b]_n .
\end{aligned} \tag{12.12}$$

Proof: Condition (12.11) implies that we can write

$$[n-b-c-1]_{n-j} = [d+e]_{n-j},$$

and as we have

$$[n-d-e-1]_j = [d+e-n+j]_j (-1)^j,$$

the product of these two factors from the right sum of (12.7) becomes

$$[n-b-c-1]_{n-j} [n-d-e-1]_j = [d+e]_{n-j} [d+e-n+j]_j (-1)^j = [d+e]_n (-1)^j.$$

The right sum therefore becomes

$$[d+e]_n \sum_{j=0}^{n} \binom{n}{j} [b+a]_j [c+a]_j [d-a]_{n-j} [e-a]_{n-j} (-1)^j.$$

But (12.11) now implies that the Pfaff–Saalschütz identity (9.1) tells us that this sum equals $(-1)^n [b+e]_n [c+e]_n$. This proves the theorem. □

Remark 12.5. If we express e using (12.11) and consider (12.12) as a polynomial identity in the variable d, then the identity between the coefficients of d^n becomes (11.2).

The following is a generalization of Dougall's formula analogous to the generalized Pfaff–Saalschütz formula (9.10).

Theorem 12.6 (Generalized Dougall's Formula). *For any $n \in \mathbb{N}$ and $a, b, c, d, e \in \mathbb{C}$ satisfying the condition*

$$p = b + c + d + e - n + 1 \in \mathbb{N}_0, \tag{12.13}$$

we have

$$\begin{aligned}
&\sum_{k=0}^{n} \binom{n}{k} [n+2a]_k [b+a]_k [c+a]_k [d+a]_k [e+a]_k \\
&\quad \times [n-2a]_{n-k} [b-a]_{n-k} [c-a]_{n-k} [d-a]_{n-k} [e-a]_{n-k} (n+2a-2k) \\
&=(-1)^n [n+2a]_{2n+1} [b+e-p]_{n-p} [b+d-p]_{n-p} [d+e-p]_{n-p} \\
&\quad \times \sum_{j=0}^{p} \binom{p}{j} [n]_j [c+a]_j [c-a]_j [b+e-n]_{p-j} [b+d-n]_{p-j} [d+e-n]_{p-j} \\
&=(-1)^n [n+2a]_{2n+1} \\
&\quad \times \sum_{j=0}^{p} \binom{p}{j} [n]_j [c+a]_j [c-a]_j [b+e-p]_{n-j} [b+d-p]_{n-j} [d+e-p]_{n-j}.
\end{aligned} \tag{12.14}$$

Proof: Theorem 12.3 yields

$$(-1)^n [n+2a]_{2n+1} S,$$

where

$$S = \sum_{j=0}^{n} \binom{n}{j} [b+a]_j [c+a]_j [n-d-e-1]_j [d-a]_{n-j} [e-a]_{n-j} [n-b-c-1]_{n-j}.$$

Using the condition (12.13), we can write the product

$$[n-d-e-1]_j [n-b-c-1]_{n-j} = (-1)^j [d+e-p]_{n-p} [j+d+e-n]_p.$$

Next we apply the Chu–Vandermonde formula (8.2) to write

$$[j+d+e-n]_p = \sum_{i=0}^{p} \binom{p}{i} [j]_i [d+e-n]_{p-i},$$

and then the product as

$$\begin{aligned}
&[n-d-e-1]_j [n-b-c-1]_{n-j} \\
&\qquad = (-1)^j [d+e-p]_{n-p} \sum_{i=0}^{p} \binom{p}{i} [j]_i [d+e-n]_{p-i} \\
&\qquad = (-1)^j \sum_{i=0}^{p} \binom{p}{i} [j]_i [d+e-p]_{n-i}.
\end{aligned}$$

Changing the order of summation we get

$$S=\sum_{i=0}^{p}\binom{p}{i}[d+e-p]_{n-i}\sum_{j=0}^{n}\binom{n}{j}[j]_i[b+a]_j[c+a]_j[d-a]_{n-j}[e-a]_{n-j}(-1)^j.$$

Using $\binom{n}{j}[j]_i=[n]_i\binom{n-i}{j-i}$ and $[x]_j=[x]_i[x-i]_{j-i}$ for $x=b+a$ and $x=c+a$, we obtain

$$\begin{aligned}S=&\sum_{i=0}^{p}\binom{p}{i}[d+e-p]_{n-i}[n]_i[b+a]_i[c+a]_i\\&\times\sum_{j=0}^{n}\binom{n-i}{j-i}[b+a-i]_{j-i}[c+a-i]_{j-i}[d-a]_{n-j}[e-a]_{n-j}(-1)^j.\end{aligned}$$

Now the condition (12.13) assures that we can apply the generalized Pfaff–Saalschütz formula (9.10) with the number p replaced by $p-i$. Therefore, the sum becomes

$$\begin{aligned}S=&\sum_{i=0}^{p}\binom{p}{i}[d+e-p]_{n-i}[n]_i[b+a]_i[c+a]_i\\&\times\sum_{\ell=0}^{p-i}\binom{p-i}{\ell}[n-i]_\ell[c+a-i]_\ell[b+d-p]_{n-i-\ell}\\&\times[b+e-p]_{n-i-\ell}(-1)^{\ell+i}.\end{aligned}$$

In this sum we can join two pairs, $[c+a]_i[c+a-i]_\ell=[c+a]_{i+\ell}$ and $[n]_i[n-i]_\ell=[n]_{i+\ell}$. Then we substitute $\ell-i$ for ℓ and interchange the order of summation to get

$$\begin{aligned}S=&\sum_{\ell=0}^{p}\binom{p}{\ell}[n]_\ell[b+d-p]_{n-\ell}[b+e-p]_{n-\ell}[c+a]_\ell(-1)^\ell\\&\times\sum_{i=0}^{\ell}\binom{\ell}{i}[d+e-p]_{n-i}[b+a]_i.\end{aligned}$$

After writing the product $[d+e-p]_{n-i}=[d+e-p]_{n-\ell}[d+e-p-n+\ell]_{\ell-i}$, we can apply the Chu–Vandermonde formula (8.2) to get for the inner sum

$$[d+e-p]_{n-\ell}[a+b+d+e-p-n+\ell]_\ell=[d+e-p]_{n-\ell}[c-a]_\ell(-1)^\ell,$$

and hence

$$S=\sum_{\ell=0}^{p}\binom{p}{\ell}[n]_\ell[b+d-p]_{n-\ell}[b+e-p]_{n-\ell}[c+a]_\ell[d+e-p]_{n-\ell}[c-a]_\ell.$$

Applying the factorization $[x-p]_{n-\ell} = [x-p]_{n-p}[x-n]_{p-\ell}$, we get the form of (12.14). $\square$

Furthermore, from L. J. Slater [Slater 66, (III.19)] we have:

Theorem 12.7. *For any $a, d \in \mathbb{C}$ and any $n \in \mathbb{N}$ we have*

$$\begin{aligned}
&\sum_{k=0}^{n} \binom{n}{k} [n+2a]_k [d+a]_k \left[d+a-\tfrac{1}{2}\right]_k [n-2d]_k [n-1+2a-2d]_k \\
&\qquad \times (n+2a-2k)[n-2a]_{n-k}[d-a]_{n-k}\left[d-a-\tfrac{1}{2}\right]_{n-k} \\
&\qquad \times [n-2d-2a]_{n-k}[n-1-2d]_{n-k} \\
&= \left(\tfrac{1}{4}\right)^n [n+2a]_{2n+1}[4d-n]_n[2d-2a]_n[2d+2a-1]_n. \qquad (12.15)
\end{aligned}$$

Proof: We apply the transformation theorem Theorem 12.3 with the arguments (a, b, c, d, e) replaced by $(a, d, d-\frac{1}{2}, n-2d-a, n-1+a-2d)$. Then we get

$$\begin{aligned}
&(-1)^n [n+2a]_{2n+1} \sum_{j=0}^{n} \binom{n}{j} [d+a]_j \left[d+a-\tfrac{1}{2}\right]_j \\
&\qquad \times [4d-n]_j [n-2d-2a]_{n-j}[n-1-2d]_{n-j}[n-2d-\tfrac{1}{2}]_{n-j}. \qquad (12.16)
\end{aligned}$$

Now we can apply Theorem 10.7 with the arguments (a, b) replaced with $(2n-1-4d, 2d+2a)$ and obtain for the sum (reversed)

$$\left(\tfrac{1}{4}\right)^n [2n-1-4d]_n[2d-2a-1]_n[2d-2a]_n. \qquad (12.17)$$

Changing the signs of the first factor, we obtain the result in the theorem. $\square$

There are three formulas to be derived from Theorems 10.7 and 10.8 with the aid of the following lemma.

Lemma 12.8. *For any $a, b \in \mathbb{C}$ we have*

$$\begin{aligned}
&\sum_{k=0}^{\lfloor \frac{n}{2} \rfloor} \binom{\lfloor \frac{n}{2} \rfloor}{k} \left[\lceil \tfrac{n}{2} \rceil - \tfrac{1}{2}\right]_k \left[\lfloor \tfrac{n}{2} \rfloor - \tfrac{1}{2}\right]_{\lfloor \frac{n}{2} \rfloor - k} \left[\tfrac{n}{2}+a\right]_k \left[\tfrac{n-1}{2}+a\right]_k \\
&\qquad \times \left[\tfrac{n}{2}-a\right]_{\lceil \frac{n}{2} \rceil - k} \left[\tfrac{n-1}{2}-a\right]_{\lfloor \frac{n}{2} \rfloor - k} \left[\tfrac{b+a}{2}\right]_k \left[\tfrac{b+a-1}{2}\right]_k \\
&\qquad \times \left[\tfrac{b-a}{2}\right]_{\lceil \frac{n}{2} \rceil - k} \left[\tfrac{b-a-1}{2}\right]_{\lfloor \frac{n}{2} \rfloor - k} (n+2a-4k) \\
&= \left[\lfloor \tfrac{n}{2} \rfloor - \tfrac{1}{2}\right]_{\lfloor \frac{n}{2} \rfloor} \left[\tfrac{n}{2}+a\right]_{n+1} \left(\left[\tfrac{n-1}{2}+a\right]_n + (-1)^n \left[\tfrac{n-1}{2}+b\right]_n\right),
\end{aligned}$$

and

$$\sum_{k=0}^{\lfloor \frac{n}{2} \rfloor} \binom{\lfloor \frac{n}{2} \rfloor}{k} \left[\lceil \tfrac{n}{2} \rceil - \tfrac{1}{2}\right]_k \left[\lfloor \tfrac{n}{2} \rfloor + \tfrac{1}{2}\right]_{\lfloor \frac{n}{2} \rfloor - k} \left[\tfrac{n}{2} + a\right]_k \left[\tfrac{n-1}{2} + a\right]_k$$
$$\times \left[\tfrac{n+1}{2} - a\right]_{\lceil \frac{n}{2} \rceil - k} \left[\tfrac{n}{2} - a\right]_{\lfloor \frac{n}{2} \rfloor - k} \left[\tfrac{b+a-2}{2}\right]_k \left[\tfrac{b+a-1}{2}\right]_k$$
$$\times \left[\tfrac{b-a}{2}\right]_{\lceil \frac{n}{2} \rceil - k} \left[\tfrac{b-a-1}{2}\right]_{\lfloor \frac{n}{2} \rfloor - k} (n + 2a - 4k - 1)$$
$$= \frac{\left[\lfloor \frac{n}{2} \rfloor + \frac{1}{2}\right]_{\lfloor \frac{n}{2} \rfloor} \left[\frac{n-1}{2} + a\right]_{n+1}}{2(n+1)(a+b)} \left(\left[\tfrac{n}{2} + a\right]_{n+1} + (-1)^n \left[\tfrac{n}{2} + b\right]_{n+1} \right).$$

Proof: The formulas are obtained by adding and subtracting the formulas (10.4) and (10.5) and then applying the expansions (5.11)–(5.21). □

Theorem 12.9. *For any $a, b \in \mathbb{C}$ and any $n \in \mathbb{N}$ we have*

$$\sum_{k=0}^{n} \binom{n}{k} \left[n - \tfrac{1}{2}\right]_k \left[n - \tfrac{1}{2}\right]_{n-k} [n+a]_k \left[n - \tfrac{1}{2} + a\right]_k [n-a]_{n-k}$$
$$\times \left[n - \tfrac{1}{2} - a\right]_{n-k} \left[\tfrac{b+a}{2}\right]_k \left[\tfrac{b+a-1}{2}\right]_k \left[\tfrac{b-a}{2}\right]_{n-k}$$
$$\times \left[\tfrac{b-a-1}{2}\right]_{n-k} (n + a - 2k)$$
$$= \frac{1}{2} \left[n - \tfrac{1}{2}\right]_n [n+a]_{2n+1} \left(\left[n - \tfrac{1}{2} + a\right]_{2n} + \left[n - \tfrac{1}{2} + b\right]_{2n} \right), \quad (12.18)$$

$$\sum_{k=0}^{n} \binom{n}{k} \left[n - \tfrac{1}{2}\right]_k \left[n + \tfrac{1}{2}\right]_{n-k} \left[n + a + \tfrac{1}{2}\right]_k [n+a]_k [n-a]_{n-k}$$
$$\times \left[n - a - \tfrac{1}{2}\right]_{n-k} \left[\frac{b+a-\frac{3}{2}}{2}\right]_k \left[\frac{b+a-\frac{1}{2}}{2}\right]_k \left[\frac{b-a-\frac{1}{2}}{2}\right]_{n-k}$$
$$\times \left[\frac{b-a-\frac{3}{2}}{2}\right]_{n-k} (n + a - 2k)$$
$$= \frac{\left[n - \frac{1}{2}\right]_n [n+a]_{2n+1}}{a + b + \frac{1}{2}} \left(\left[n + a + \tfrac{1}{2}\right]_{2n+1} + [n+b]_{2n+1} \right), \quad (12.19)$$

$$\sum_{k=0}^{n} \binom{n}{k} \left[n + \tfrac{1}{2}\right]_k \left[n + \tfrac{1}{2}\right]_{n-k} \left[n + \tfrac{1}{2} + a\right]_k [n+a]_k [n-a]_{n-k}$$
$$\times \left[n + \tfrac{1}{2} - a\right]_{n-k} \left[\tfrac{b+a-2}{2}\right]_k \left[\tfrac{b+a-1}{2}\right]_k \left[\tfrac{b-a-2}{2}\right]_{n-k}$$
$$\times \left[\tfrac{b-a-1}{2}\right]_{n-k} (n + a - 2k)$$
$$= \frac{\left[n + \frac{1}{2}\right]_n [n+a]_{2n+1}}{(n+1)(a+b)(a-b)} \left(\left[n + \tfrac{1}{2} + a\right]_{2n+2} - \left[n + \tfrac{1}{2} + b\right]_{2n+2} \right). \quad (12.20)$$

Proof: The formulas follow immediately from Lemma 12.8 upon replacing n by $2n$ and $2n + 1$. □

12.3 Type II(8,8,1)

There is a single formula known to us of this type.

Theorem 12.10. *For any $a, b, c \in \mathbb{C}$ and $n, p \in \mathbb{N}_0$, if*

$$p = 2(a + b + c + 1 - n) + \tfrac{1}{2},$$

then we have the following balanced formula

$$\begin{aligned}
\sum_{k=0}^{n} \binom{n}{k} & \left[n+\tfrac{1}{2}\right]_k \left[n-\tfrac{1}{2}\right]_{n-k} \left[a+\tfrac{1}{2}\right]_k \left[b+\tfrac{1}{2}\right]_k \left[c+\tfrac{1}{2}\right]_k [a]_k[b]_k[c]_k \\
& \times [a]_{n-k}[b]_{n-k}[c]_{n-k} \left[a-\tfrac{1}{2}\right]_{n-k} \left[b-\tfrac{1}{2}\right]_{n-k} \left[c-\tfrac{1}{2}\right]_{n-k} \\
= (-1)^p & \left(\tfrac{1}{64}\right)^n [2n]_n [2a]_n[2b]_n[2c]_n \\
& \times \left[2a+\tfrac{1}{2}-p\right]_{n-p} \left[2b+\tfrac{1}{2}-p\right]_{n-p} \left[2c+\tfrac{1}{2}-p\right]_{n-p} \\
& \times \sum_{i=0}^{p} \binom{p}{i} [n]_i[2c+1]_i[2c-n]_i \\
& \qquad \times \left[p-2c-\tfrac{3}{2}\right]_{p-i} \left[2a+\tfrac{1}{2}-n\right]_{p-i} \left[2b+\tfrac{1}{2}-n\right]_{p-i}.
\end{aligned}$$

Proof: We apply Theorem (10.1) to $2n+1$, $2a+1$, $2b+1$, and $2c+1$, and the observation that the corresponding symmetric sum of type II$(4,4,-1)$ must vanish. Adding the two sums we can apply (5.15) and (5.16) to the binomial coefficient and split the 6 factorials into two each. □

For $p = 0$ the formula has a much nicer right side:

Corollary 12.11. *For any $a, b, c \in \mathbb{C}$ and $n \in \mathbb{N}_0$, if*

$$a + b + c + 1 - n + \tfrac{1}{4} = 0,$$

then we have the balanced formula

$$\begin{aligned}
\sum_{k=0}^{n} \binom{n}{k} & \left[n+\tfrac{1}{2}\right]_k \left[n-\tfrac{1}{2}\right]_{n-k} \left[a+\tfrac{1}{2}\right]_k \left[b+\tfrac{1}{2}\right]_k \left[c+\tfrac{1}{2}\right]_k [a]_k[b]_k[c]_k \\
& \times [a]_{n-k}[b]_{n-k}[c]_{n-k} \left[a-\tfrac{1}{2}\right]_{n-k} \left[b-\tfrac{1}{2}\right]_{n-k} \left[c-\tfrac{1}{2}\right]_{n-k} \\
= & \left(\tfrac{1}{4096}\right)^n [2n]_n [4a+1]_{2n}[4b+1]_{2n}[4c+1]_{2n}.
\end{aligned}$$

Proof: Obvious. □

12.4 Type II(p, p, z)

The general formula for this family is the following, cf. [Gould 72b, (1.53)]:

Theorem 12.12. *For $p, n \in \mathbb{N}$, $y \in \mathbb{C}$ and $q, r \in \mathbb{N}_0$ satisfying $r, q < p$ and $r + q \leq p$, with $\rho = e^{\frac{2\pi i}{p}}$ and $z = y^p$, we have the equivalent identities*

$$\sum_{k=0}^{n} \binom{pn+r+q}{pk+r} z^k = \frac{y^{-r}}{p} \sum_{j=0}^{p-1} (1+\rho^j y)^{pn+r+q} \rho^{-jr}, \tag{12.21}$$

$$\sum_{k=0}^{n} \binom{n}{k} \prod_{j=1}^{q} \left[n+\tfrac{j}{p}\right]_k \prod_{j=1}^{p-q-1} \left[n-\tfrac{j}{p}\right]_k \prod_{j=1}^{r} \left[n+\tfrac{j}{p}\right]_{n-k} \prod_{j=1}^{p-r-1} \left[n-\tfrac{j}{p}\right]_{n-k} z^k$$
$$= \frac{\prod_{j=1}^{r} \left[n+\frac{j}{p}\right]_n \prod_{j=1}^{p-r-1} \left[n-\frac{j}{p}\right]_n}{p\binom{pn+r+q}{r} y^r} \sum_{j=0}^{p-1} (1+\rho^j y)^{pn+r+q} \rho^{-jr}. \tag{12.22}$$

Proof: Consider the sum on the right sides of (12.21)–(12.22), and apply the binomial formula to the power:

$$\begin{aligned} S &= \sum_{j=0}^{p-1} (1+\rho^j y)^{pn+r+q} \rho^{-jr} = \sum_{j=0}^{p-1} \sum_{i=0}^{pn+r+q} \binom{pn+r+q}{i} \rho^{ij} y^i \rho^{-rj} \\ &= \sum_{i=0}^{pn+r+q} \binom{pn+r+q}{i} y^i \sum_{j=0}^{p-1} \rho^{(i-r)j}. \end{aligned} \tag{12.23}$$

As we have

$$\sum_{j=0}^{p-1} \rho^{(i-r)j} = \begin{cases} p & \text{if } i \equiv r\ (p), \\ 0 & \text{if } i \not\equiv r\ (p), \end{cases}$$

we can sum only the nonzero terms by changing the summation variable in (12.23) to k, where $i = r + kp$, to get

$$S = \sum_{k=0}^{n} \binom{pn+r+q}{pk+r} y^{pk+r} p = p y^r \sum_{k=0}^{n} \binom{pn+r+q}{pk+r} z^k. \tag{12.24}$$

This yields (12.21).

To obtain (12.22), it only remains to write the binomial coefficient in the appropriate way, i.e.,

$$
\begin{aligned}
\binom{pn+r+q}{pk+r} &= \frac{[pn+r+q]_{pk+r}}{[pk+r]_{pk+r}} = \frac{[pn+r+q]_r\,[pn+q]_{pk}}{[pk+r]_{pk}\,[r]_r} \\
&= \binom{pn+r+q}{r}\frac{\prod_{j=0}^{p-1}[pn+q-j,p]_k}{\prod_{j=0}^{p-1}[pk+r-j,p]_k} \\
&= \binom{pn+r+q}{r}\binom{n}{k}\frac{\prod_{j=1}^{q}\left[n+\frac{j}{p}\right]_k\prod_{j=1}^{p-q-1}\left[n-\frac{j}{p}\right]_k}{\prod_{j=1}^{r}\left[k+\frac{j}{p}\right]_k\prod_{j=1}^{p-r-1}\left[k-\frac{j}{p}\right]_k} \\
&= \frac{\binom{pn+r+q}{r}}{\prod_{j=1}^{r}\left[n+\frac{j}{p}\right]_n\prod_{j=1}^{p-r-1}\left[n-\frac{j}{p}\right]_n}\binom{n}{k}\prod_{j=1}^{q}\left[n+\frac{j}{p}\right]_k \\
&\quad\times\prod_{j=1}^{p-q-1}\left[n-\frac{j}{p}\right]_k\prod_{j=1}^{r}\left[n+\frac{j}{p}\right]_{n-k}\prod_{j=1}^{p-r-1}\left[n-\frac{j}{p}\right]_{n-k}.
\end{aligned}
\tag{12.25}
$$

Substitution of (12.25) into (12.24) yields (12.22). □

The special choice of $p = 2$ gives (8.15) for $(r, q) = (0, 0)$, (8.16) for $(r, q) = (0, 1)$, and (8.17) for $(r, q) = (1, 1)$.

Chapter 13

Zeilberger's Algorithm

In 1990 Doron Zeilberger, working with Herbert Wilf and later Marco Petkovšek, (see [Zeilberger 90], [Wilf and Zeilberger 90], [Koornwinder 91], [Graham et al. 94, p. 239 ff.], [Petkovšek et al. 96]) used Gosper's algorithm to prove formulas of the following form, where a is independent of n and k, and the limits are natural in the sense that the terms vanish outside the interval of summation:

$$T(a,n) = \sum_{k=0}^{n} t(a,n,k).$$

It is required that each quotient

$$q_t(a,n,k) = \frac{t(a,n,k+1)}{t(a,n,k)}$$

be rational not only as a function of k, but as a function of n as well, and furthermore, that the quotients

$$r_t(a,n,k) = \frac{t(a,n+1,k)}{t(a,n,k)} \tag{13.1}$$

be rational as functions of both k and n.

They proved that the method works in all cases where the terms are *proper*, by which they meant that the terms consist of products of polynomials in n and k, powers to the degrees n and k, and factorials of the form $[a+pk+qn]_m$, where $a \in \mathbb{C}$ and $p, q, m \in \mathbb{Z}$, and, of course, that the terms vanish outside the interval of summation.

In the simplest cases the method not only provides us with a tool to prove known—or guessed—formulas, but also enables the discovery of one.

The aim is to find a difference equation satisfied by the sums $T(a,n)$ with coefficients that are rational in n, i.e., to find an equation

$$\beta_0(n)T(a,n) + \beta_1(n)T(a,n+1) + \cdots + \beta_\ell(n)T(a,n+\ell) = 0.$$

To do that we seek to sum the corresponding terms

$$\beta_0(n)t(a,n,k) + \beta_1(n)t(a,n+1,k) + \cdots + \beta_\ell(n)t(a,n+\ell,k)$$

by Gosper's algorithm on the form $\Delta S(a,n,k)$, because with the terms, the sums $S(a,n,k)$ must also vanish outside a finite interval. Hence, we have

$$\sum_{k=-\infty}^{\infty} \Delta S(a,n,k) = \sum_{k=-\infty}^{\infty} S(a,n,k+1) - \sum_{k=-\infty}^{\infty} S(a,n,k) = 0.$$

Of course we want the order of the difference equation to be as small as possible, so we start with $\ell = 1$ and try to determine the coefficients such that Gosper's algorithm works. If not, we try again with $\ell = 2$, etc. If we succeed with a first-order equation, we can always find the sum. Otherwise, we can use the difference equation to prove a known formula, unless we are able to solve the equation. But this is often enough not the case, cf. Chapter 4.

To make the ideas more transparent, let us consider the case $\ell = 1$. Then we handle the quotient of

$$s(a,n,k) = \beta_0(n)t(a,n,k) + \beta_1(n)t(a,n+1,k), \tag{13.2}$$

namely,

$$q_s(a,n,k) = \frac{\beta_0(n)t(a,n,k+1) + \beta_1(n)t(a,n+1,k+1)}{\beta_0(n)t(a,n,k) + \beta_1(n)t(a,n+1,k)}.$$

We can apply (13.1) to write this quotient as

$$q_s(a,n,k) = \frac{\beta_0(n) + \beta_1(n)r_t(a,n,k+1)}{\beta_0(n) + \beta_1(n)r_t(a,n,k)} \cdot \frac{t(a,n,k+1)}{t(a,n,k)}.$$

If the function r_t behaves nicely, we can rewrite this quotient as

$$q_s(a,n,k) = \frac{f(a,n,k+1)}{f(a,n,k)} \cdot \frac{g(a,n,k)}{h(a,n,k+1)},$$

with f, g, and h polynomials in k with coefficients which are rational functions of n. If this is the case, we proceed with Gosper, forming

$$\begin{aligned} G(a,n,k) &= g(a,n,k) - h(a,n,k), \\ H(a,n,k) &= g(a,n,k) + h(a,n,k), \end{aligned}$$

to estimate the degree d of the polynomial s in k which solves

$$f(a,n,k) = s(a,n,k+1)g(a,n,k) - s(a,n,k)h(a,n,k). \tag{13.3}$$

This polynomial identity must be solved in the $d+1$ unknown coefficients in s and the $\ell+1=2$ unknown coefficients in f as rational functions of n. As soon as the existence of a solution is verified, we are satisfied with the two functions, β, or even less, namely their ratio,

$$\frac{\beta_0(n)}{\beta_1(n)}, \tag{13.4}$$

because we then find immediately the desired sum as

$$T(a,n) = T(a,0)\prod_{\nu=0}^{n-1} -\frac{\beta_0(\nu)}{\beta_1(\nu)}.$$

13.1 A Simple Example of Zeilberger's Algorithm

Let us consider the sum (in standard form)

$$T(n) = \sum_k \binom{n}{k} [a]_k [b]_{n-k} x^k,$$

and look at the term (13.2) with $\ell = 1$. It can be written as

$$\begin{aligned} s(n,k) &= \beta_0(n)\binom{n}{k}[a]_k[b]_{n-k}x^k + \beta_1(n)\binom{n+1}{k}[a]_k[b]_{n+1-k}x^k \\ &= \left(\frac{n+1-k}{n+1}\beta_0(n) + (b-n+k)\beta_1(n)\right)\binom{n+1}{k}[a]_k[b]_{n-k}x^k. \end{aligned} \tag{13.5}$$

The quotient hence becomes

$$\begin{aligned} q_s(n,k) = &\frac{(n-k)\beta_0(n) + (b-n+k+1)(n+1)\beta_1(n)}{(n+1-k)\beta_0(n) + (b-n+k)(n+1)\beta_1(n)} \\ &\times \frac{(n+1-k)(a-k)x}{(-1-k)(n-1-b-k)}, \end{aligned} \tag{13.6}$$

and it is obvious how to obtain the polynomials as

$$\begin{aligned} f(n,k) &= (n+1-k)\beta_0(n) + (b-n+k)(n+1)\beta_1(n), \\ g(n,k) &= xk^2 - x(n+1+a)k + xa(n+1), \\ h(n,k) &= k^2 + (b-n)k, \\ G(n,k) &= (x-1)k^2 - (x(n+1+a) - n + b)k + xa(n+1), \\ H(n,k) &= (x+1)k^2 - (x(n+1+a) + n - b)k + xa(n+1). \end{aligned}$$

Now, if $x \neq 1$, the degrees of G and H are both 2 while the degree of f is 1, leaving us with no solution. Nevertheless, if we allow $\ell = 2$, there is a solution, cf. (8.79). The following solution was found by use of a Maple program due to D. Zeilberger:

$$T(n+2)+((1+x)(n+1)-xa-b)\,T(n+1)+x(n-a-b)(n+1)T(n) = 0. \tag{13.7}$$

Recent versions of Maple include a substantial implementation of Zeilberger's method.

But, for $x = 1$, we get the reduction

$$G(n,k) = (-1-a-b)k + a(n+1).$$

Hence, the degree from (6.10) becomes 0, so we just have one rational function of n, $\alpha(n)$.

The equation (13.3) takes the simple form

$$f(n,k) = \alpha(n)G(n,k),$$

or

$$(n+1-k)\beta_0(n)+(b-n+k)(n+1)\beta_1(n) = \alpha(n)((-1-a-b)k+a(n+1)),$$

or, comparing coefficients of the two polynomials, we get the pair of equations

$$-\beta_0(n) + (n+1)\beta_1(n) = \alpha(n)(-1-a-b),$$
$$(n+1)\beta_0(n) + (b-n)(n+1)\beta_1(n) = \alpha(n)a(n+1)),$$

yielding the desired fraction (13.4) as

$$\frac{\beta_0(n)}{\beta_1(n)} = -(a+b-n).$$

This gives the solution

$$T(n) = [a+b]_n,$$

which is the Chu–Vandermonde formula (8.2).

13.2 A Less Simple Example of Zeilberger's Algorithm

Let us consider the sum (in standard form)

$$T(n) = \sum_k \binom{n}{k} [a]_k [b]_k [c]_{n-k} [n-1-a-b-c]_{n-k} (-1)^k,$$

and look at the term (13.2) with $\ell = 1$. It can be written as

$$\begin{aligned}
s(n,k) &= \beta_0(n)\binom{n}{k}[a]_k[b]_k[c]_{n-k}[n-1-a-b-c]_{n-k}(-1)^k \\
&\quad + \beta_1(n)\binom{n+1}{k}[a]_k[b]_k[c]_{n+1-k}[n-a-b-c]_{n+1-k}(-1)^k \\
&= \left(\frac{n+1-k}{n+1}\beta_0(n) + (c-n+k)(n-a-b-c)\beta_1(n)\right) \\
&\quad \times \binom{n+1}{k}[a]_k[b]_k[c]_{n-k}[n-1-a-b-c]_{n-k}(-1)^k.
\end{aligned}$$

Hence, the quotient becomes

$$\begin{aligned}
q_s(n,k) =& \frac{(n-k)\beta_0(n) + (c-n+k+1)(n-a-b-c)(n+1)\beta_1(n)}{(n+1-k)\beta_0(n) + (c-n+k)(n-a-b-c)(n+1)\beta_1(n)} \\
&\times \frac{(n+1-k)(a-k)(b-k)}{(-1-k)(n-1-c-k)(a+b+c-k)},
\end{aligned}$$

and it is obvious how to obtain the polynomials as

$$\begin{aligned}
f(n,k) =& (n+1-k)\beta_0(n) + (c-n+k)(n-a-b-c)(n+1)\beta_1(n), \\
g(n,k) =& k^3 - (n+1+a+b)k^2 + ((a+b)(n+1)+ab)k - (n+1)ab, \\
h(n,k) =& k^3 - (n+1+a+b)k^2 + (n-c)(a+b+c+1)k, \\
G(n,k) =& ((c+1)(a+b+c-n)+ab)k - (n+1)ab, \\
H(n,k) =& 2k^3 - 2(n+1+a+b)k^2 \\
&+ ((n+1)(a+b) + ab + (n-c)(a+b+c+1))k - (n+1)ab.
\end{aligned}$$

We get the degree from (6.10) to be 0, so we have just one rational function of n, $\alpha(n)$.

Equation (13.3) takes the simple form $f(n,k) = \alpha(n)G(n,k)$, or

$$\begin{aligned}
(n+1-k)\beta_0(n) &+ (c-n+k)(n-a-b-c)(n+1)\beta_1(n) \\
&= \alpha(n)(((c+1)(a+b+c-n)+ab)k - (n+1)ab),
\end{aligned}$$

or, comparing coefficients of the two polynomials, we get a pair of equations

$$\begin{aligned}
-\beta_0(n) + (n-a-b-c)(n+1)\beta_1(n) &= \alpha(n)((c+1)(a+b+c-n)+ab), \\
(n+1)\beta_0(n) + (c-n)(n-a-b-c)(n+1)\beta_1(n) &= -\alpha(n)(n+1)ab,
\end{aligned}$$

having the desired fraction (13.4) to be

$$\frac{\beta_0(n)}{\beta_1(n)} = (a+c-n)(b+c-n),$$

which yields the solution

$$T(n) = [a+c]_n[b+c]_n(-1)^n, \tag{13.8}$$

the well-known Pfaff–Saalschütz formula, cf. (9.1).

13.3 Sporadic Formulas of Types II(2,2,z)

Besides (13.8) there are many formulas with the limit of summation as the only free parameter. In each case, the transformation group gives up to six equivalent formulas with different factors. We choose the biggest of the six factors in the following presentation of such formulas.

The Factor 2

Besides the theorems of Gauss and Bailey, there are a few other formulas known with a factor of 2.

$$\sum_{k=0}^{n}\binom{n}{k}\left[4n+\tfrac{1}{2}\right]_k\left[-n-\tfrac{1}{2}\right]_{n-k}2^k = \frac{\left[-\frac{3}{8}\right]_n\left[-\frac{5}{8}\right]_n}{\left[-\frac{1}{2}\right]_n}(-16)^n, \tag{13.9}$$

$$\sum_{k=0}^{n}\binom{n}{k}\left[4n+\tfrac{5}{2}\right]_k\left[-n-\tfrac{3}{2}\right]_{n-k}2^k = \frac{\left[-\frac{7}{8}\right]_n\left[-\frac{9}{8}\right]_n}{\left[-\frac{3}{2}\right]_n}(-16)^n. \tag{13.10}$$

The Factor 4

The following seven formulas are known:

$$\sum_{k=0}^{n}\binom{n}{k}\left[3n+\tfrac{1}{2}\right]_k\left[-\tfrac{1}{2}\right]_{n-k}4^k = \left[-\tfrac{1}{2}\right]_n(-27)^n, \tag{13.11}$$

$$\sum_{k=0}^{n}\binom{n}{k}\left[3n+\tfrac{3}{2}\right]_k\left[-\tfrac{1}{2}\right]_{n-k}4^k = \frac{\left[-\frac{5}{6}\right]_n\left[-\frac{7}{6}\right]_n}{\left[-\frac{3}{2}\right]_n}(-27)^n = \frac{[6n+1,2]_{2n}}{[2n]_n}, \tag{13.12}$$

$$\sum_{k=0}^{n}\binom{n}{k}\left[3n+\tfrac{1}{2}\right]_k[-3n-1]_{n-k}4^k = \frac{\left[-\frac{5}{6}\right]_n\left[-\frac{1}{2}\right]_n}{\left[-\frac{2}{3}\right]_n}(-16)^n, \tag{13.13}$$

$$\sum_{k=0}^{n}\binom{n}{k}\left[3n+\tfrac{3}{2}\right]_k[-3n-3]_{n-k}4^k = \frac{\left[-\frac{5}{6}\right]_n\left[-\frac{3}{2}\right]_n}{\left[-\frac{5}{3}\right]_n}(-16)^n, \tag{13.14}$$

$$\sum_{k=0}^{n}\binom{n}{k}\left[-\tfrac{1}{2}\right]_k\left[\tfrac{3n}{2}\right]_{n-k}4^k = \left[\tfrac{3n}{2}\right]_n\frac{2(-1)^n+1}{3}, \tag{13.15}$$

$$\sum_{k=0}^{n}\binom{n}{k}\left[-\tfrac{1}{2}\right]_k\left[\tfrac{3n+1}{2}\right]_{n-k}4^k=\begin{cases}0 & \text{if } n \text{ is odd},\\ [n]_m^2\left(\frac{27}{16}\right)^m & \text{if } n=2m.\end{cases} \tag{13.16}$$

The formulas (13.11) and (13.16) are due to Ira M. Gessel, see [Gessel and Stanton 82, (5.22) and (5.25)].

The formula (13.15) allows a slight generalization for even values of the limit of summation.

$$\sum_{k=0}^{2n}\binom{2n}{k}\left[-\tfrac{1}{2}-p\right]_k[3n+p]_{2n-k}4^k=\frac{[3n+p]_{2n+p}}{[p]_p},\quad p=0,1,2. \tag{13.17}$$

The formula (13.17) for $p=0$ is due to Ira M. Gessel [Gessel 95a, (28.1a)], and the formula for $p=1$ is from Gould's table [Gould 72b, (7.8)].

The Factor 5

The following four are known:

$$\sum_{k=0}^{n}\binom{n}{k}\left[5n+\tfrac{1}{2}\right]_k\left[n-\tfrac{1}{2}\right]_{n-k}5^k=\frac{\left[-\frac{1}{2}\right]_n\left[-\frac{3}{10}\right]_n\left[-\frac{7}{10}\right]_n}{\left[-\frac{2}{5}\right]_n\left[-\frac{3}{5}\right]_n}(-64)^n, \tag{13.18}$$

$$\sum_{k=0}^{n}\binom{n}{k}\left[5n+\tfrac{3}{2}\right]_k\left[n-\tfrac{1}{2}\right]_{n-k}5^k=\frac{\left[-\frac{1}{2}\right]_n\left[-\frac{9}{10}\right]_n\left[-\frac{11}{10}\right]_n}{\left[-\frac{4}{5}\right]_n\left[-\frac{6}{5}\right]_n}(-64)^n, \tag{13.19}$$

$$\sum_{k=0}^{n}\binom{n}{k}\left[5n+\tfrac{5}{2}\right]_k\left[n+\tfrac{1}{2}\right]_{n-k}5^k=\frac{\left[-\frac{3}{2}\right]_n\left[-\frac{7}{10}\right]_n\left[-\frac{13}{10}\right]_n}{\left[-\frac{7}{5}\right]_n\left[-\frac{8}{5}\right]_n}(-64)^n, \tag{13.20}$$

$$\sum_{k=0}^{n}\binom{n}{k}\left[5n+\tfrac{7}{2}\right]_k\left[n+\tfrac{1}{2}\right]_{n-k}5^k=\frac{\left[-\frac{3}{2}\right]_n\left[-\frac{9}{10}\right]_n\left[-\frac{11}{10}\right]_n}{\left[-\frac{6}{5}\right]_n\left[-\frac{9}{5}\right]_n}(-64)^n. \tag{13.21}$$

The Factor 9

The following eighteen are known:

$$\sum_{k=0}^{n}\binom{n}{k}\left[2n-\tfrac{8}{3}\right]_k[-2n+3]_{n-k}9^k=\frac{[-2n+3]_n}{\left[\frac{2}{3}\right]_n}\times\left(\left[\tfrac{5}{6}\right]_n+\left(\tfrac{3}{4}-\tfrac{n}{3}\right)\left[\tfrac{1}{2}\right]_{n-1}\right)(-4)^n\quad n>1, \tag{13.22}$$

$$\sum_{k=0}^{n}\binom{n}{k}\left[2n-\tfrac{4}{3}\right]_k[-2n+1]_{n-k}\,9^k=\frac{[3n-2]_n}{\left[\frac{1}{3}\right]_n}\left(\left[\tfrac{1}{6}\right]_n+\tfrac{1}{2}\left[\tfrac{1}{2}\right]_n\right)4^n, \tag{13.23}$$

$$\sum_{k=0}^{n}\binom{n}{k}\left[2n-\tfrac{2}{3}\right]_k[-2n]_{n-k}\,9^k=\frac{[3n-1]_n}{\left[-\frac{1}{3}\right]_n}\left(\left[-\tfrac{1}{6}\right]_n+\tfrac{1}{2}\left[-\tfrac{1}{2}\right]_n\right)4^n, \tag{13.24}$$

$$\sum_{k=0}^{n}\binom{n}{k}\left[2n+\tfrac{1}{3}\right]_k[-2n-1]_{n-k}\,9^k=\left[-\tfrac{2}{3}\right]_n(-27)^n, \tag{13.25}$$

$$\sum_{k=0}^{n}\binom{n}{k}\left[2n+\tfrac{2}{3}\right]_k[-2n-2]_{n-k}\,9^k=\frac{\left[-\frac{5}{6}\right]_n\left[-\frac{4}{3}\right]_n}{\left[-\frac{3}{2}\right]_n}(-27)^n, \tag{13.26}$$

$$\sum_{k=0}^{n}\binom{n}{k}\left[2n+\tfrac{4}{3}\right]_k[-2n-3]_{n-k}\,9^k$$
$$=6\frac{\left[-\frac{5}{3}\right]_n}{[2n+2]_{n+2}}\left(\left[-\tfrac{1}{6}\right]_{n+1}-\left[-\tfrac{1}{2}\right]_{n+1}\right)108^n, \tag{13.27}$$

$$\sum_{k=0}^{n}\binom{n}{k}\left[3n+\tfrac{1}{2}\right]_k\left[n-\tfrac{1}{2}\right]_{n-k}\,9^k=\left[-\tfrac{1}{2}\right]_n(-64)^n, \tag{13.28}$$

$$\sum_{k=0}^{n}\binom{n}{k}\left[3n+\tfrac{3}{2}\right]_k\left[n+\tfrac{1}{2}\right]_{n-k}\,9^k=\frac{\left[-\frac{3}{2}\right]_n\left[-\frac{5}{6}\right]_n\left[-\frac{7}{6}\right]_n}{\left[-\frac{4}{3}\right]_n\left[-\frac{5}{3}\right]_n}(-64)^n, \tag{13.29}$$

$$\sum_{k=0}^{n}\binom{n}{k}\left[3n+\tfrac{1}{4}\right]_k\left[n-\tfrac{1}{4}\right]_{n-k}\,9^k=\frac{\left[-\frac{3}{4}\right]_n\left[-\frac{5}{12}\right]_n}{\left[-\frac{2}{3}\right]_n}(-64)^n, \tag{13.30}$$

$$\sum_{k=0}^{n}\binom{n}{k}\left[3n+\tfrac{3}{4}\right]_k\left[n+\tfrac{1}{4}\right]_{n-k}\,9^k=\frac{\left[-\frac{5}{4}\right]_n\left[-\frac{7}{12}\right]_n}{\left[-\frac{4}{3}\right]_n}(-64)^n, \tag{13.31}$$

$$\sum_{k=0}^{n}\binom{n}{k}\left[3n+\tfrac{5}{4}\right]_k\left[n-\tfrac{1}{4}\right]_{n-k}\,9^k=\frac{\left[-\frac{3}{4}\right]_n\left[-\frac{13}{12}\right]_n}{\left[-\frac{4}{3}\right]_n}(-64)^n, \tag{13.32}$$

$$\sum_{k=0}^{n}\binom{n}{k}\left[3n+\tfrac{7}{4}\right]_k\left[n+\tfrac{1}{4}\right]_{n-k}\,9^k=\frac{\left[-\frac{5}{4}\right]_n\left[-\frac{11}{12}\right]_n}{\left[-\frac{5}{3}\right]_n}(-64)^n, \tag{13.33}$$

$$\sum_{k=0}^{n}\binom{n}{k}\left[\tfrac{n}{2}-\tfrac{1}{3}\right]_k\left[-\tfrac{n}{2}-1\right]_{n-k}\,9^k$$
$$=\begin{cases}0 & \text{for } n \text{ odd},\\ \left[-\frac{1}{2}\right]_m\left[-\frac{2}{3}\right]_m(-108)^m & \text{for } n=2m,\end{cases} \tag{13.34}$$

$$\sum_{k=0}^{n}\binom{n}{k}\left[\tfrac{n}{2}-\tfrac{1}{3}\right]_k\left[-\tfrac{n}{2}\right]_{n-k}9^k$$
$$=\begin{cases}\left[\frac{1}{3}\right]_n\cdot(-1)^{m+1}3^{n+m} & \text{for } n=2m+1,\\ \left(\left[\frac{1}{3}\right]_n+2\left[-\frac{1}{2}\right]_m\left[-\frac{2}{3}\right]_m 4^m\right)\cdot(-1)^m 3^{n+m-1} & \text{for } n=2m,\end{cases}\tag{13.35}$$

$$\sum_{k=0}^{n}\binom{n}{k}\left[\tfrac{n}{2}-\tfrac{1}{6}\right]_k\left[-\tfrac{n}{2}-\tfrac{1}{2}\right]_{n-k}9^k=\left[-\tfrac{2}{3}\right]_n(-1)^{\lfloor\frac{n}{2}\rfloor}3^{n+\lfloor\frac{n}{2}\rfloor},\tag{13.36}$$

$$\sum_{k=0}^{n}\binom{n}{k}\left[\tfrac{n}{2}+\tfrac{1}{6}\right]_k\left[-\tfrac{n}{2}-\tfrac{3}{2}\right]_{n-k}9^k$$
$$=\begin{cases}\frac{1}{n+1}\left[-\frac{1}{3}\right]_{n+1}(-1)^{m+1}3^{n+m+1} & \text{for } n=2m,\\ \frac{1}{n+1}\Big(\left[-\frac{1}{3}\right]_{n+1}(-1)^{m+1}3^{n+m+1} & \\ \quad -\frac{1}{3}\left[-\frac{1}{2}\right]_{m+1}\left[-\frac{2}{3}\right]_{m+1}(-108)^{m+1}\Big) & \text{for } n=2m+1.\end{cases}\tag{13.37}$$

For n even, the formulas (13.22)–(13.26), and (13.34)–(13.37) are found in [Gessel and Stanton 82] as 3.10, 3.9, 3.8, 5.23, 3.7, 5.24, 3.13, 3.12, and 3.14, respectively. The formulas (13.30) and (13.32) are mentioned as conjectures of Gosper in [Gessel and Stanton 82], formulas 6.5 and 6.6, respectively.

Two similar formulas are known only for n even, namely,

$$\sum_{k=0}^{n}\binom{n}{k}\left[\tfrac{n}{2}+\tfrac{1}{3}\right]_k\left[-\tfrac{n}{2}-2\right]_{n-k}9^k=\frac{\left[-\frac{5}{3}\right]_n(-1)^{\frac{n}{2}}3^{n+\frac{n}{2}+1}}{n+2},\tag{13.38}$$

$$\sum_{k=0}^{n}\binom{n}{k}\left[\tfrac{n}{2}+\tfrac{2}{3}\right]_k\left[-\tfrac{n}{2}-3\right]_{n-k}9^k=\tfrac{5}{(n+2)(n+4)}\cdot\left[-\tfrac{7}{3}\right]_n(-1)^{\frac{n}{2}}3^{n+\frac{n}{2}+1}.\tag{13.39}$$

The formulas (13.38) and (13.39) are found in [Gessel and Stanton 82, (3.15) and (3.16)].

13.4 Sporadic Formulas of Types II(4,4,z)

A few formulas are known with $z\neq\pm1$, all of them proved using the Zeilberger method by I. M. Gessel [Gessel 95a]. They have z equal to 4, $\frac{32}{27}$, $\frac{2}{27}$, $-\frac{9}{16}$, and -8, respectively.

The formulas with $z = -8$ are in [Gessel 95a] as numbers 11.1a, 11.2a, 21.1a, 21.3a, and 30.3a.

$$\sum_{k=0}^{n} \binom{n}{k} [n+a]_k \left[a - \tfrac{1}{2}\right]_k [-n-1]_{n-k} [n-2a]_{n-k} (2a+2n-3k) 8^k$$
$$= [2a]_{n+1} [-a-1]_n (-4)^n, \tag{13.40}$$

$$\sum_{k=0}^{n} \binom{n}{k} [a-1]_k \left[n - a + \tfrac{1}{2}\right]_k [n+1-2a]_{n-k} [2a-n-2]_{n-k}$$
$$\times (2n-3k) 8^k = 0, \tag{13.41}$$

$$\sum_{k=0}^{n} \binom{n}{k} [n-b]_k \left[-b - \tfrac{1}{2}\right]_k [-n-1]_{n-k} [2b+n]_{n-k}$$
$$\times (2(n-b)-3k) 8^k = [b-1]_n [-2b]_{n+1}, \tag{13.42}$$

$$\sum_{k=0}^{n} \binom{n}{k} [n-b]_k \left[-b - \tfrac{1}{2}\right]_k [-n-1]_{n-k} [2b+n]_{n-k}$$
$$\times (2(n-b)-3k) 8^k = [b-1]_n [-2b]_{n+1}, \tag{13.43}$$

$$\sum_{k=0}^{n} \binom{n}{k} [-a-1/2]_k [n+a]_k [n+2a]_{n-k} [-1-2a-n]_{n-k}$$
$$\times (2n-3k) 8^k = 0. \tag{13.44}$$

The formula with $z = -\frac{9}{16}$ is in [Gessel 95a, (28.7a)].

$$\sum_{k=0}^{n} \binom{n}{k} \left[-\tfrac{1}{2}\right]_k \left[2n + \tfrac{1}{2}\right]_k \left[-\tfrac{n}{2} - 1\right]_{n-k}$$
$$\times \left[-\tfrac{n}{2} - \tfrac{1}{2}\right]_{n-k} (3(4n+1) - 10k) \left(\frac{9}{16}\right)^k = 3[4n+1]_{2n+1} \left(\frac{27}{256}\right)^n. \tag{13.45}$$

The formulas with $z = \frac{2}{27}$ are in [Gessel 95a] as numbers 30.4a, 30.5a, 30.6a, and 30.7a.

$$\sum_{k=0}^{n} \binom{n}{k} \left[-\tfrac{2}{3}\right]_k [-2n-2]_k \left[2n + \tfrac{5}{6}\right]_{n-k} \left[\tfrac{1}{3} + n\right]_{n-k}$$
$$\times (2n+5+10k) \left(-\frac{2}{27}\right)^k = 5 \frac{[n+1/3]_n \, [2n+5/6]_{2n}}{[2n+1]_{n+1}} 4^n, \tag{13.46}$$

$$\sum_{k=0}^{n} \binom{n}{k} \left[-\tfrac{1}{6}\right]_k [-2n]_k \left[n-\tfrac{2}{3}\right]_{n-k} \left[-\tfrac{2}{3}+2n\right]_{n-k} (n+5k)\Big(-\frac{2}{27}\Big)^k$$
$$= \frac{n\,[2n-2/3]_{2n}\,[n-2/3]_n}{[2n]_n} 4^n, \tag{13.47}$$

$$\sum_{k=0}^{n} \binom{n}{k} \left[-\tfrac{1}{3}\right]_k [-2n-1]_k \left[n-\tfrac{1}{3}\right]_{n-k} \left[\tfrac{1}{6}+2n\right]_{n-k} (2n+1+10k)\Big(-\frac{2}{27}\Big)^k$$
$$= \frac{[n-1/3]_n\,[2n+1/6]_{2n}}{[2n]_n} 4^n, \tag{13.48}$$

$$\sum_{k=0}^{n} \binom{n}{k} \left[-\tfrac{5}{6}\right]_k [-2-2n]_k \left[n+\tfrac{2}{3}\right]_{n-k} \left[2n+\tfrac{2}{3}\right]_{n-k} (n+4+5k)\Big(-\frac{2}{27}\Big)^k$$
$$= \frac{\left[n+\tfrac{2}{3}\right]_n \left[2n+\tfrac{2}{3}\right]_{2n}}{[2n+1]_{n+1}} 4^{n+1}. \tag{13.49}$$

The formulas with $z=\frac{32}{27}$ are in [Gessel 95a] as number 30.10a.

$$\sum_{k=0}^{n} \binom{n}{k} \left[-\tfrac{1}{3}\right]_k \left[n-\tfrac{1}{2}\right]_k \left[n-\tfrac{1}{3}\right]_{n-k} [-3n]_{n-k} (4n+5k)\Big(\frac{32}{27}\Big)^k$$
$$= 3\frac{\left[n-\tfrac{1}{3}\right]_n \left[2n-\tfrac{5}{6}\right]_{2n}}{[3n-1]_{n-1}} (-16)^n, \tag{13.50}$$

$$\sum_{k=0}^{n} \binom{n}{k} \left[-\tfrac{1}{3}\right]_k \left[n+\tfrac{1}{2}\right]_k \left[n-\tfrac{1}{3}\right]_{n-k} [-3n-2]_{n-k} (4n+2+5k)\Big(\frac{32}{27}\Big)^k$$
$$= 12\frac{\left[n-\tfrac{1}{3}\right]_n \left[2n+\tfrac{1}{6}\right]_{2n+1}}{[3n+1]_n} (-16)^n. \tag{13.51}$$

The formulas with $z=4$ are in [Gessel 95a] as numbers 22.1a, 27.3a, 29.1a, 29.2a, 29.4a, and 29.5a.

$$\sum_{k=0}^{n} \binom{n}{k} \left[b+\tfrac{n}{2}\right]_k \left[b+\tfrac{n-1}{2}\right]_k [n-b]_{n-k}[-2b]_{n-k}(2b+n-3k)(-4)^k$$
$$= -[b-1]_n[-2b]_{n+1}, \tag{13.52}$$

$$\sum_{k=0}^{n} \binom{n}{k} \left[n-\tfrac{1}{2}\right]_k [3n+a]_k[a]_{n-k}[-a]_{n-k}(3n+a-3k)(-4)^k$$
$$= (-1)^n a[3n+a]_{2n}, \tag{13.53}$$

$$\sum_{k=0}^{n} \binom{n}{k} \left[b - \tfrac{1}{2}\right]_k [n+b]_k [n-b]_{n-k} [n-2b]_{n-k}$$
$$\times (2(n+b) - 3k)(-4)^k = -2[2b-1]_n [-b]_{n+1}, \tag{13.54}$$

$$\sum_{k=0}^{n} \binom{n}{k} \left[n - \tfrac{1}{2}\right]_k [2b]_k [b]_{n-k} [-n]_{n-k} (2n-3k)(-4)^k = -(-1)^n [2n]_{n+1} [b]_n, \tag{13.55}$$

$$\sum_{k=0}^{n} \binom{n}{k} \left[\tfrac{n-3}{2} - a\right]_k \left[\tfrac{n-2}{2} - a\right]_k [a+n+1]_{n-k} [2(a+1)]_{n-k}$$
$$\times (2(a+1) - n + 3k)(-4)^k = -2[2a+1]_n [-a-1]_{n+1}, \tag{13.56}$$

$$\sum_{k=0}^{n} \binom{n}{k} \left[n - \tfrac{1}{2}\right]_k [a-1]_k [-n]_{n-k} \left[-\tfrac{1}{2} + \tfrac{a}{2} - n\right]_{n-k} (4n-3k)(-4)^k$$
$$= [2n]_{n+1} \left[\tfrac{a}{2} - 1\right]_n (-4)^n, \tag{13.57}$$

$$\sum_{k=0}^{n} \binom{n}{k} \left[n + \tfrac{1}{2}\right]_k [a-1]_k [-n-1]_{n-k} \left[\tfrac{a}{2} - \tfrac{3}{2} - n\right]_{n-k} (4n+2-3k)(-4)^k$$
$$= 2\,[2n+1]_{n+1} \left[\tfrac{a}{2} - 1\right]_n (-4)^n. \tag{13.58}$$

Chapter 14

Sums of Types III–IV

14.1 The Abel, Hagen–Rothe, Cauchy, and Jensen Formulas

The classical formulas of type III are the Hagen–Rothe formula and the Jensen formula. The Hagen–Rothe formula is a generalization of the Chu–Vandermonde identity:

$$\sum_{k=0}^{n} \frac{x}{x+kz}\binom{x+kz}{k}\frac{y}{y+(n-k)z}\binom{y+(n-k)z}{n-k} \tag{14.1}$$

$$= \frac{x+y}{x+y+nz}\binom{x+y+nz}{n}. \tag{14.2}$$

It was established in 1793 by H. A. Rothe [Rothe 93] and quoted by J. G. Hagen [Hagen 91] in 1891. It can be found in H. W. Gould [Gould 72b, (3.142) or (3.146)], J. Kaucký [Kaucký 75, (6.4)], J. Riordan [Riordan 68, p. 169] or in [Andersen 89] and [Andersen and Larsen 94].

The Jensen formula generalizes the binomial formula:

$$\sum_{k=0}^{n} \binom{n}{k}(x+kz)^{k-1}(y-kz)^{n-k-1} = \frac{x+y-nz}{x(y-nz)}(x+y)^{n-1}. \tag{14.3}$$

It was established in 1902 by J. L. W. V. Jensen [Jensen 02, (10)]; see H. W. Gould [Gould 72b, (1.125)], J. Kaucký [Kaucký 75, (6.6.5)], or [Andersen 89], [Andersen and Larsen 94].

Both formulas are valid for any complex number z. To treat these two formulas as one, we introduce

$$\begin{aligned} S &= S(x,y,z,d;p,q,n) \\ &= \sum_{k=0}^{n} \binom{n}{k} [x+kz-pd,d]_{k-p}[y+(n-k)z-qd,d]_{n-k-q}, \end{aligned} \tag{14.4}$$

where $x, y, z, d \in \mathbb{C}$, $p, q \in \mathbb{Z}$, and $n \in \mathbb{N}_0$. This is in analogy with J. Riordan's treatment of N. H. Abel's identities [Riordan 68, pp. 18–23] [Abel 39].

Note that by changing the direction of summation we get the corresponding sum with x exchanged with y and p with q:

$$S(y,x,z,d;q,p,n) = S(x,y,z,d;p,q,n). \tag{14.5}$$

For $p = q = 1$ and $d = 1$, after division by $[n]_n$, the sum becomes equal to the left side of (14.2), while for $d = 0$, after the substitution of $y - nz$ for y, it is equal to the left side of (14.3).

For $d = 0$ the addition of the terms kz was studied by A.-L. Cauchy [Cauchy 87] in 1826 in the case of $p = q = 0$, and the same year by Abel [Abel 39] for $p = 1, q = 0$; see also [Gould 72b, (1.124)], J. Riordan [Riordan 68, p. 8], and [Andersen 89, p. 20]. Jensen was the first one to study the case $p = q = 1$ in 1902 [Jensen 02].

For $d = 1$ the story took the opposite direction. The case $p = q = 1$ was studied already in 1793 by H. A. Rothe [Rothe 93] and again by J. G. Hagen in 1891 [Hagen 91], while study of the cases $p = q = 0$ and $p = 1, q = 0$ was done by Jensen in 1902 [Jensen 02, formulas 17 and 18]. See also H. W. Gould, [Gould 72b, formula 3.144] and J. Kaucký, [Kaucký 75, (6.4.2)] and [Andersen 89, p. 19].

We will follow the approach in [Andersen and Larsen 94]. We will allow d to be arbitrary, and then prove the formulas for $(p,q) = (0,0)$, then for $(p,q) = (1,0)$ (and from (14.5) also for $(p,q) = (0,1)$), and eventually for $(p,q) = (1,1)$. For technical reasons we will replace y with $y+nz$ in (14.4).

Theorem 14.1 (Generalized Cauchy–Jensen Identity). *For $x, y, z, d \in \mathbb{C}$ and $n \in \mathbb{N}_0$*

$$\sum_{k=0}^{n} \binom{n}{k} [x+kz,d]_k[y-kz,d]_{n-k} = \sum_{j=0}^{n} [n]_j z^j [x+y-jd,d]_{n-j}. \tag{14.6}$$

Proof: Let

$$S = S(x, y - nz, z, d; 0, 0, n) = \sum_{k=0}^{n} \binom{n}{k} [x + kz, d]_k [y - kz, d]_{n-k}. \quad (14.7)$$

We apply (8.3) to rewrite the first factor as

$$[x + kz, d]_k = \sum_{i=0}^{k} \binom{k}{i} [x, d]_i [kz, d]_{k-i}.$$

Now we can write S as a double sum

$$S = \sum_{k=0}^{n} \binom{n}{k} \sum_{i=0}^{k} \binom{k}{i} [x, d]_i [kz, d]_{k-i} [y - kz, d]_{n-k}.$$

We apply (2.1) to the last factor, and then (8.3):

$$\begin{aligned}(-1)^{n-k}[y - kz, d]_{n-k} &= [-y + kz + (n - k - 1)d, d]_{n-k} \\ &= \sum_{j=i}^{n-k+i} \binom{n-k}{j-i} [-y + (n - i - 1)d, d]_{j-i} [kz + (i - k)d, d]_{n-k-j+i}.\end{aligned}$$

We then get S written as a triple sum:

$$\begin{aligned}S = &\sum_{k=0}^{n} \binom{n}{k} \sum_{i=0}^{k} \binom{k}{i} \sum_{j=i}^{n-k+i} \binom{n-k}{j-i} [x, d]_i [kz, d]_{k-i} (-1)^{n-k} \\ &\times [-y + (n - i - 1)d, d]_{j-i} [kz + (i - k)d, d]_{n-k-j+i}.\end{aligned}$$

Now we apply (2.9) and (2.12) several times to the product of the three binomial coefficients and get

$$\begin{aligned}\binom{n}{k}\binom{k}{i}\binom{n-k}{j-i} &= \binom{n}{i}\binom{n-i}{k-i}\binom{n-k}{j-i} = \binom{n}{i}\binom{n-i}{n-k}\binom{n-k}{j-i} \\ &= \binom{n}{i}\binom{n-i}{j-i}\binom{n-j}{n-k-j+i} = \binom{n}{j}\binom{j}{i}\binom{n-j}{k-i}.\end{aligned}$$

After this, the sum (14.7) can be rearranged to

$$S = \sum_{j=0}^{n} \binom{n}{j} \sum_{i=0}^{j} \binom{j}{i} [x, d]_i [-y + (n - i - 1)d, d]_{j-i} (-1)^{n-i} T_{ji}(z, d; n), \quad (14.8)$$

where

$$T_{ji}(z,d;n) = \sum_{k=i}^{n-j+i} \binom{n-j}{k-i} [kz,d]_{k-i}[kz+(i-k)d,d]_{n-k-j+i}(-1)^{k-i}. \tag{14.9}$$

Now we apply (2.2) to the sum (14.9) and obtain

$$T_{ji}(z,d;n) = \sum_{k=i}^{n-j+i} \binom{n-j}{k-i} [kz,d]_{n-j}(-1)^{k-i}.$$

Shifting the variable of summation by i, we get

$$T_{ji}(z,d;n) = \sum_{k=0}^{n-j} \binom{n-j}{k} [(k+i)z,d]_{n-j}(-1)^{k}. \tag{14.10}$$

Now the factorial in (14.10) is a polynomial in k of the form

$$[(k+i)z,d]_{n-j} = \prod_{\ell=0}^{n-j-1} (kz+iz-\ell d) = z^{n-j}k^{n-j} + \cdots .$$

Hence, when we apply (8.10) to (14.10), we only get one term, which, as $\mathfrak{S}_n^{(n)} = 1$, becomes

$$T_{ji}(z,d;n) = z^{n-j}[n-j]_{n-j}(-1)^{n-j}. \tag{14.11}$$

When we embed (14.11) into (14.8), we get the sum

$$S = \sum_{j=0}^{n} \binom{n}{j} \sum_{i=0}^{j} \binom{j}{i} [x,d]_i[-y+(n-i-1)d,d]_{j-i}(-1)^{j-i}z^{n-j}[n-j]_{n-j}.$$

Now we can place the terms which are independent of i outside the inner sum and apply (2.12) and (2.8) outside the inner sum and (2.1) inside it to get

$$S = \sum_{j=0}^{n} [n]_{n-j}z^{n-j} \sum_{i=0}^{j} \binom{j}{i} [x,d]_i[y-(n-j)d,d]_{j-i}.$$

Then we can apply (8.3) to the inner sum to eliminate it:

$$S = \sum_{j=0}^{n} [n]_{n-j}z^{n-j}[x+y-(n-j)d,d]_j.$$

When we eventually change the direction of summation, we get

$$S = \sum_{j=0}^{n} [n]_j z^j [x + y - jd, d]_{n-j},$$

which gives (14.6). □

If we replace y with $y + nz$ in (14.6), we get the expression

$$\sum_{k=0}^{n} \binom{n}{k} [x + kz, d]_k [y + (n-k)z, d]_{n-k} = \sum_{j=0}^{n} [n]_j z^j [x + y + nz - jd, d]_{n-j}. \tag{14.12}$$

Now, let p and q be arbitrary integers again. With the help of (14.6) and (14.12) we want to prove a recursion which allows us to find a generalization of the formulas of Abel and Jensen for $p = 1$ and $q = 0$.

Proposition 14.2. *For all p, $q \in \mathbb{Z}$, $n \in \mathbb{N}$ we have*

$$\begin{aligned} S(x, y, z, d; p, q, n) &= (x - pd) S(x, y, z, d; p+1, q, n) \\ &\quad + nzS(x - d + z, y, z, d; p, q, n-1). \end{aligned} \tag{14.13}$$

Proof: We apply (2.2) with $m = 1$ to the first factor in (14.4) to write it as

$$[x + kz - pd, d]_{k-p} = (x + kz - pd)[x + kz - (p+1)d, d]_{k-(p+1)}.$$

We split the factor $(x + kz - pd)$ into two parts, $(x - pd)$ and kz, by which S is split into two sums, and apply (2.8) to the second sum:

$$\begin{aligned} &S(x, y, z, d; p, q, n) \\ &= \sum_{k=0}^{n} \binom{n}{k} (x - pd)[x + kz - (p+1)d, d]_{k-p-1} [y + (n-k)z - qd, d]_{n-k-q} \\ &\quad + \sum_{k=1}^{n} \binom{n}{k} kz[x + kz - (p+1)d, d]_{k-p-1} [y + (n-k)z - qd, d]_{n-k-q} \\ &= (x - pd) S(x, y, z, d; p+1, q, n) \\ &\quad + nz \sum_{k=1}^{n} \binom{n-1}{k-1} [x - d + kz - pd, d]_{k-1-p} [y + (n-k)z - qd, d]_{n-k-q} \\ &= (x - pd) S(x, y, z, d; p+1, q, n) + nzS(x - d + z, y, z, d; p, q, n-1). \end{aligned}$$

□

With this we have a two-dimensional recursion formula in the variables p and n relating the three pairs (p,n), $(p+1,n)$, and $(p,n-1)$. As we know the sum in (14.12) for $p=q=0$ and all n, we can now use the recursion formula.

We choose $p=q=0$ in (14.13) and by applying (14.12) we get

$$\begin{aligned}
&S(x,y,z,d;1,0,n)\\
&\quad= \frac{1}{x}\left(S(x,y,z,d;0,0,n) - nzS(x-d+z,y,z,d;0,0,n-1)\right)\\
&\quad= \frac{1}{x}\Bigg(\sum_{j=0}^{n}[n]_j z^j [x+y+nz-jd,d]_{n-j}\\
&\qquad\quad - nz\sum_{j=0}^{n-1}[n-1]_j z^j[x-d+z+y+(n-1)z-jd,d]_{n-1-j}\Bigg)\\
&\quad= \frac{1}{x}\Bigg(\sum_{j=0}^{n}[n]_j z^j [x+y+nz-jd,d]_{n-j}\\
&\qquad\quad - \sum_{j=0}^{n-1}[n]_{j+1} z^{j+1}[x+y+nz-(j+1)d,d]_{n-(j+1)}\Bigg)\\
&\quad= \frac{1}{x}\Bigg(\sum_{j=0}^{n}[n]_j z^j [x+y+nz-jd,d]_{n-j}\\
&\qquad\quad - \sum_{j=1}^{n}[n]_j z^j[x+y+nz-jd,d]_{n-j}\Bigg)\\
&\quad= \frac{1}{x}[x+y+nz,d]_n.
\end{aligned}$$

We have proved the formula:

Theorem 14.3 (Generalized Abel–Jensen Identity I). *For $n \in \mathbb{N}_0$ and $x,y,z,d \in \mathbb{C}$,*

$$\sum_{k=0}^{n}\binom{n}{k}[x+kz-d,d]_{k-1}[y+(n-k)z,d]_{n-k} = \frac{1}{x}[x+y+nz,d]_n. \tag{14.14}$$

and the formula we get by exchanging y with $y-nz$:

Corollary 14.4 (Generalized Abel–Jensen Identity II). *For $n \in \mathbb{N}_0$ and $x,y,z,d \in \mathbb{C}$,*

$$\sum_{k=0}^{n}\binom{n}{k}[x+kz-d,d]_{k-1}[y-kz,d]_{n-k} = \frac{1}{x}[x+y,d]_n. \tag{14.15}$$

For $d = 0$, (14.15) becomes the formula of Abel from 1826 [Abel 39], and for $d = 1$ after division by $[n]_n$, it becomes the formula of Jensen from 1902 [Jensen 02, (17)].

The surprise is that for $p = 1$ and $q = 0$ the sum is reduced to a single term, while the sum for $p = q = 0$ only is rewritten as another sum. Even more surprising is the fact that a single term also is sufficient for $p = q = 1$.

For this purpose we will apply the symmetry (14.5) to evaluate the sum for $p = 0$ and $q = 1$. Then we will apply the following recursion in p, q, and n.

Proposition 14.5. *For all $p, q \in \mathbb{Z}$, $n \in \mathbb{N}$,*

$$\begin{aligned} S(x, y, z, d; p, q, n) = S(x + z - d, y, z, d; p - 1, q, n - 1) \\ + S(x, y + z - d, z, d; p, q - 1, n - 1). \end{aligned} \tag{14.16}$$

Proof: Using (2.7) we split the sum (14.4) into

$$\begin{aligned} S &= \sum_{k=0}^{n} \binom{n}{k} [x + kz - pd, d]_{k-p} [y + (n-k)z - qd, d]_{n-k-q} \\ &= \sum_{k} \binom{n-1}{k-1} [x + kz - pd, d]_{k-p} [y + (n-k)z - qd, d]_{n-k-q} \\ &\quad + \sum_{k} \binom{n-1}{k} [x + kz - pd, d]_{k-p} [y + (n-k)z - qd, d]_{n-k-q} \\ &= \sum_{k} \binom{n-1}{k} [x + z - d + kz - (p-1)d, d]_{k-(p-1)} \\ &\qquad \times [y + (n-1-k)z - qd, d]_{n-1-k-q} \\ &\quad + \sum_{k} \binom{n-1}{k} [x + kz - pd, d]_{k-p} \\ &\qquad \times [y + z - d + (n-1-k)z - (q-1)d, d]_{n-1-k-(q-1)} \\ &= S(x + z - d, y, z, d; p - 1, q, n - 1) \\ &\quad + S(x, y + z - d, z, d; p, q - 1, n - 1). \end{aligned}$$

□

When we put $p = q = 1$ in (14.16) and apply (14.14) and its symmetric companion to the first and second of the two sums, respectively, we get

$$\begin{aligned} S(x, y, z, d; 1, 1, n) = S(x + z - d, y, z, d; 0, 1, n - 1) \\ + S(x, y + z - d, z, d; 1, 0, n - 1) \end{aligned}$$

$$= \frac{1}{y}[x+z-d+y+(n-1)z,d]_{n-1}$$
$$+\frac{1}{x}[y+z-d+x+(n-1)z,d]_{n-1}$$
$$= \frac{x+y}{xy}[x+y+nz-d,d]_{n-1}.$$

Hence, we have proved:

Theorem 14.6 (Generalized Hagen–Rothe–Jensen Identity I). *For $n \in \mathbb{N}_0$ and $x, y, z, d \in \mathbb{C}$ we have the formula*

$$\sum_{k=0}^{n} \binom{n}{k} [x+kz-d,d]_{k-1}[y+(n-k)z-d,d]_{n-k-1}$$
$$= \frac{x+y}{xy}[x+y+nz-d,d]_{n-1}. \tag{14.17}$$

As above we can write y for $y+nz$ and get the equivalent formula.

Corollary 14.7 (Generalized Hagen–Rothe–Jensen Identity II). *For $n \in \mathbb{N}_0$ and $x, y, z, d \in \mathbb{C}$ we have the formula*

$$\sum_{k=0}^{n} \binom{n}{k} [x+kz-d,d]_{k-1}[y-kz-d,d]_{n-k-1} = \frac{x+y-nz}{x(y-nz)}[x+y-d,d]_{n-1}. \tag{14.18}$$

For $d=1$, (14.17) is identical to (14.2) after multiplication by $[n]_n$, while $d=0$ makes (14.18) identical to (14.3).

14.2 A Polynomial Identity

In 1995 [Andersen and Larsen 95] we posed the problem of proving the following:

If $x \in \mathbb{C}$ and $n \in \mathbb{N}$, then

$$(-4)^n \sum_{j=0}^{n} \binom{x+\frac{1}{2}}{j}\binom{n-1-x}{2n-j} = \binom{2n}{n}\sum_{j=0}^{n}\binom{x+j}{2j}\binom{x-j}{2n-2j}; \tag{14.19}$$

if furthermore $m \in \mathbb{N}$ with $0 \le m \le 2n$, then

$$(-4)^n \sum_{j=0}^{n} \binom{x+\frac{1}{2}}{j}\binom{n-1-x}{2n-j} = \binom{2n}{n}\sum_{j=-\lfloor\frac{m}{2}\rfloor}^{n-\lfloor\frac{m}{2}\rfloor}\binom{x+j}{2j+m}$$
$$\times\binom{x-j}{2n-m-2j}. \tag{14.20}$$

Proof: We give the proof of (14.19). The proof of (14.20) is similar, but with some technical complications.

The two sides of (14.19) represent polynomials in x of degree $2n$, so it is sufficient to establish the identity for $2n+1$ different values of x. The points $x = 0, 1, \ldots, n-1$ will be proved zeros of the two polynomials. The left-hand side is zero for $x = 0, 1, \ldots, n-1$ because $0 \leq n-1-x < n \leq 2n-j$. The right-hand side is zero for $x = 0, 1, \ldots, n-1$ because we have $x+j \geq 0$, so to get a nonzero term, we must have $x+j \geq 2j$, i.e., $x - j \geq 0$. Hence, the other factor can only be nonzero if $x - j \geq 2n - 2j$, i.e., $x + j \geq 2n$, and hence $2x \geq 2n$, contrary to the assumption.

Now consider $x = y - \frac{1}{2}$, $y = 0, 1, \ldots, n$. The polynomials will be shown to take the same constant value for all of these points.

The left-hand side becomes

$$(-4)^n \sum_{j=0}^{n} \binom{y}{j} \binom{n - \frac{1}{2} - y}{2n - j} = (-4)^n \sum_{j=0}^{n} \binom{y}{j} \frac{[n - \frac{1}{2} - y]_{2n-j}}{[2n - j]_{2n-j}}.$$

As long as $y \leq n$, the limit of summation becomes y. So we get

$$\frac{(-4)^n [n - \frac{1}{2} - y]_{2n-y}}{[2n]_{2n}} \sum_{j=0}^{y} \binom{y}{j} [-n - \tfrac{1}{2}]_{y-j} [2n]_j.$$

But this is a Chu–Vandermonde sum, (8.3) or (B.4), so we get

$$\begin{aligned}
&\frac{(-4)^n \left[n - \frac{1}{2} - y\right]_{2n-y}}{[2n]_{2n}} \left[n - \tfrac{1}{2}\right]_y \\
&\quad = \frac{(-4)^n \left[n - \frac{1}{2}\right]_{2n}}{[2n]_{2n}} = \frac{(-4)^n \left[n - \frac{1}{2}\right]_n \left[-\frac{1}{2}\right]_n}{[2n]_{2n}} \\
&\quad = \frac{[2n-1, 2]_n^2 [2n, 2]_n^2}{[2n]_{2n} [n]_n^2} \left(\tfrac{1}{2}\right)^{2n} = \frac{[2n]_{2n}}{[n]_n^2} \left(\tfrac{1}{2}\right)^{2n} = \binom{2n}{n} \left(\tfrac{1}{2}\right)^{2n}.
\end{aligned} \tag{14.21}$$

The right-hand side of (14.19) can be written for $x = y - \frac{1}{2}$ as

$$\frac{1}{[n]_n^2} \sum_{k \equiv 0(2)} \binom{2n}{k} \left[y - \tfrac{1}{2} + \tfrac{k}{2}\right]_k \left[y - \tfrac{1}{2} - \tfrac{k}{2}\right]_{2n-k}. \tag{14.22}$$

If in the form of the sum (14.22) we consider an odd k, $0 < k < 2n$, then the terms are zero for $y = 0, 1, \ldots, n$. If $k \leq y - \frac{1}{2} + \frac{k}{2}$ and $2n - k \leq y - \frac{1}{2} - \frac{k}{2}$, then we get $2n \leq 2y - 1$, i.e., $n < y$. And if we have $y - \frac{1}{2} + \frac{k}{2} < 0$ and $y - \frac{1}{2} + \frac{k}{2} \geq 2n - k$, then we must have $2n < 0$, and if $y - \frac{1}{2} - \frac{k}{2} < 0$ and then $y - \frac{1}{2} + \frac{k}{2} \geq k$, then we must have $y - \frac{1}{2} - \frac{k}{2} \geq 0$, too. Eventually, if both $y-\frac{1}{2}-\frac{k}{2}<0$ and $y-\frac{1}{2}+\frac{k}{2}<0$, then we get $2y-1<0$, so that we must have $y=0$. But then $-\frac{1}{2}+\frac{k}{2}<0$, so that $k<1$, which is impossible for an integer $k>0$. So one of the factors must vanish.

This means that we might as well take the sum for all indices $k = 0,\ldots,2n$. The sum in (14.22) equals

$$\frac{1}{[n]_n^2}\sum_{k=0}^{2n}\binom{2n}{k}\left[y-\tfrac{1}{2}+\tfrac{k}{2}\right]_k\left[y-\tfrac{1}{2}-\tfrac{k}{2}\right]_{2n-k}. \tag{14.23}$$

We can now apply Jensen's formula (14.6). With $n=2n$, $x=y=y-\frac{1}{2}$, $d=1$, and $z=\frac{1}{2}$ substituted in (14.23), we rewrite the sum as

$$\frac{1}{[n]_n^2}\sum_{j=0}^{2n}[2n]_j\left(\tfrac{1}{2}\right)^j[2y-1-j]_{2n-j}. \tag{14.24}$$

Then rewrite

$$[2y-1-j]_{2n-j}=(-1)^j[2n-2y]_{2n-j}=(-1)^j[2n-j]_{2n-j}\binom{2n-2y}{2n-j}. \tag{14.25}$$

Now two factors combine, $[2n]_j[2n-j]_{2n-j}=[2n]_{2n}$, so that after substitution of (14.25) into (14.24), the sum takes the form

$$\begin{aligned}\frac{[2n]_{2n}}{[n]_n^2}\sum_{j=0}^{2n}\binom{2n-2y}{2n-j}\left(-\tfrac{1}{2}\right)^j &= \binom{2n}{n}\left(\tfrac{1}{2}\right)^{2y}\sum_{j=0}^{2n}\binom{2n-2y}{j-2y}\left(-\tfrac{1}{2}\right)^{j-2y}\\ &=\binom{2n}{n}\left(\tfrac{1}{2}\right)^{2n},\end{aligned}$$

where we have applied the binomial theorem, (7.1).

This is the same result as in (14.21), therefore the identity (14.19) is established. □

The generalization (14.20) can be proved the same way without serious trouble.

14.3 Joseph Sinyor and Ted Speevak's Problem

The problem is to prove a combinatorial identity conjectured by Joseph Sinyor and Ted Speevak in discrete mathematics [Sinyor and Speevak 01].

In their formulation, for $m > \ell \geq 0 \leq j \leq 2\ell + 1$, the double sum

$$\sum_{\ell' \leq \ell} \sum_{j' \leq j} \frac{1}{(m-\ell')} \binom{m+\ell'-j'}{2\ell'-j'+1} \binom{m-\ell'+j'-1}{j'} \times \binom{m-\ell'}{2(\ell-\ell')-(j-j')} \binom{m-\ell'}{(j-j')} \tag{14.26}$$

is conjectured to be equal to

$$\frac{1}{2(m-\ell)} \binom{2m}{2\ell+1} \binom{2\ell+1}{j} = \frac{1}{2\ell+1} \binom{2m}{2\ell} \binom{2\ell+1}{j}. \tag{14.27}$$

Sinyor and Speevak support the conjecture with proofs of four special cases. In the case of $j = 0$, they reduce the formula to the Pfaff–Saalschütz identity (9.1). In the case of $j = 1$, they give a sophisticated combinatorial argument applying partition, and the cases of $j = 2\ell$ and $j = 2\ell + 1$, are reduced to the cases $j = 1$ and $j = 0$, respectively. See Remarks 14.8, 14.10, and 14.11 below.

Of course, both sides of the formula have been computed for many integral values of j and ℓ and complex values of m, and no counterexamples have been found. But recall the formula

$$\sum_j \sum_k \binom{n}{j-k} \binom{m}{j+k} = 2^{n+m-1},$$

where the summation is over all integer values of j and k such that the binomial coefficients do not vanish. This is true for all natural numbers m and n, which are not both zero, but fails for $n = m = 0$. So, despite convincing experimental evidence, a proof is needed.

Changing the Problem to a Polynomial Identity

We prefer to change the names of the variables, x for m, n for j, k for j', and i for ℓ'. Then (14.26) looks like

$$\sum_{i \leq \ell} \sum_{k \leq n} \frac{1}{x-i} \binom{x+i-k}{2i-k+1} \binom{x-i+k-1}{k} \binom{x-i}{2(\ell-i)-(n-k)} \binom{x-i}{n-k}. \tag{14.28}$$

And (14.27) looks like

$$\frac{1}{2(x-\ell)} \binom{2x}{2\ell+1} \binom{2\ell+1}{n} = \frac{1}{2\ell+1} \binom{2x}{2\ell} \binom{2\ell+1}{n}. \tag{14.29}$$

Now multiply (14.28) and (14.29) by $n!(2\ell+1-n)!$, which is a "constant" (i.e., independent of the summation indices i and k), to get

$$\sum_i \sum_k \binom{2\ell+1-n}{2i-k+1}\binom{n}{k}[x+i-k]_{2i-k+1}[x-i+k-1]_k \times [x-i-1]_{2\ell-1-2i+k-n}[x-i]_{n-k}. \qquad (14.30)$$

At the same time, multiplication of the result (14.29) gives the simplification that only a single factorial is left:

$$[2x]_{2\ell}. \qquad (14.31)$$

In the forms (14.30) and (14.31) the number ℓ only appears with the factor 2. Hence, it is tempting to write $m = 2\ell$ and ask whether the formula is true for all $m \geq 0$, not only even values. Doing so, we ask for a proof of the equality of the sum

$$\sum_i \sum_k \binom{m+1-n}{2i-k+1}\binom{n}{k}[x+i-k]_{2i-k+1}[x-i+k-1]_k \times [x-i-1]_{m-1-2i+k-n}[x-i]_{n-k} \qquad (14.32)$$

with the factorial

$$[2x]_m. \qquad (14.33)$$

The sum in (14.32) is a polynomial in x of degree m, so we may as well establish the identity of (14.32) with (14.33) for all $x \in \mathbb{C}$, but we must keep the condition, $0 \leq n \leq m+1$.

We know the zeros of the polynomial (14.33), namely $\frac{j}{2}$, $j = 0, \ldots, m-1$. We will prove that these are also zeros of the polynomial (14.32). When this is done we only need to establish that the two polynomials coincide in one additional point.

Fortunately, the factorials in (14.32) combine using (2.2):

$$[x+i-k]_{2i-k+1}[x-i-1]_{m-1-2i+k-n} = [x+i-k]_{m-n}$$
$$[x-i+k-1]_k[x-i]_{n-k} = (x-i)[x-i+k-1]_{n-1}.$$

Hence, we can rewrite (14.32) as

$$\sum_i \sum_k \binom{m+1-n}{2i-k+1}\binom{n}{k}(x-i)[x+i-k]_{m-n}[x-i+k-1]_{n-1}. \qquad (14.34)$$

Rewriting the Double Sum as Two Single Sums

Now it is time to change variables in (14.34). We write $i = j + k$ and change the summation indices to j and k and get

$$\sum_j [x+j]_{m-n}[x-1-j]_{n-1} \sum_k \binom{m+1-n}{2j+k+1}\binom{n}{k}(x-j-k). \quad (14.35)$$

The inner sum can be written as the difference of two sums

$$\sum_k \binom{m+1-n}{2j+k+1}\binom{n}{k}(x-j) - \sum_k \binom{m+1-n}{2j+k+1}\binom{n}{k}k. \quad (14.36)$$

Both are Chu–Vandermonde sums, cf. (8.1), i.e.,

$$\begin{aligned}\sum_k \binom{m+1-n}{2j+k+1}\binom{n}{k}(x-j) &= (x-j)\sum_k \binom{m+1-n}{m-2j-n-k}\binom{n}{k}\\ &= (x-j)\binom{m+1}{m-2j-n},\\ \sum_k \binom{m+1-n}{2j+k+1}\binom{n}{k}k &= n\sum_k \binom{m+1-n}{m-2j-n-k}\binom{n-1}{k-1}\\ &= n\binom{m}{m-2j-n-1}.\end{aligned}$$

This means that the (14.35) can be written as the difference of two polynomials:

$$\begin{aligned}&\sum_j \binom{m+1}{m-2j-n}(x-j)[x+j]_{m-n}[x-j-1]_{n-1}\\ &\quad - n\sum_j \binom{m}{m-2j-n-1}[x+j]_{m-n}[x-j-1]_{n-1},\\ &= \sum_j \binom{m+1}{2j+n+1}[x+j]_{m-n}[x-j]_n\\ &\quad - \sum_j \binom{m}{2j+n+1}[x+j]_{m-n}\, n[x-j-1]_{n-1}, \quad (14.37)\end{aligned}$$

where we have applied (2.2) and (2.12).

Now by (2.4),

$$n[x-j-1]_{n-1} = [x-j]_n - [x-j-1]_n$$

and by (2.7),

$$\binom{m+1}{2j+n+1} = \binom{m}{2j+n+1} + \binom{m}{2j+n},$$

so we can write the difference (14.37) as a sum, namely

$$\sum_j \binom{m}{2j+n} [x+j]_{m-n}[x-j]_n + \sum_j \binom{m}{2j+n+1} [x+j]_{m-n}[x-j-1]_n. \tag{14.38}$$

The Integral Zeros of (14.38)

The polynomial (14.38) is zero for integral values of x, $x = 0, \ldots, \left[\frac{m-1}{2}\right]$ because it is two sums of zeros. The numbers $x+j$ and $x-j$ cannot both be negative, and if $x+j \geq m-n$ and $x-j \geq n$, then $2x \geq m$. If $x+j \geq m-n$, but $x-j(-1) < 0$, then $2j+n(+1) > m-x+x = m$, so that the binomial coefficient is zero. If $x+j < 0$ and $x-j(-1) \geq n$, then $2j+n(+1) < x-x = 0$, so again the binomial coefficient is zero.

The Evaluation of the Polynomial (14.38) for x=–1

For $x = -1$ we can compute the value of the polynomial. First, if $n \leq m$, then in the first sum of (14.38) only the term corresponding to $j = 0$ is different from zero, and in the second sum, only the terms corresponding to $j = 0, -1$ are different from zero. Hence, the sum of the sums is the sum of the three terms

$$\begin{aligned}
\binom{m}{n}[-1]_{m-n}[-1]_n &= (-1)^m [m]_n [m-n]_{m-n} \\
&= (-1)^m [m]_m, && (14.39) \\
\binom{m}{n+1}[-1]_{m-n}[-2]_n &= (-1)^m [m]_{n+1} [m-n]_{m-n} \\
&= (m-n)(-1)^m [m]_m, && (14.40) \\
\binom{m}{n-1}[-2]_{m-n}[-1]_n &= (-1)^m [m]_{n-1} [m-n+1]_{m-n+1}\, n \\
&= n(-1)^m [m]_m, && (14.41)
\end{aligned}$$

where we have applied (2.1) and (2.2).

Summing (14.39)–(14.41) gives

$$(1+m-n+n)(-1)^m[m]_m = (-1)^m[m+1]_{m+1} = -[-1]_{m+1} = [-2]_m, \tag{14.42}$$

which is the expected value according to (14.33).

If $n = m+1$, then the first sum of (14.38) is zero and the second sum can have only the term of $j = -1$ different from zero. So the value of (14.38) equals the value of (14.41), i.e., $(m+1)(-1)^m[m]_m = [-2]_m$, as in (14.42).

The Non-integral Zeros of (14.38)

In order to find the remaining zeros of the polynomial (14.38), we change variables in the sums to $k = 2j+n$ and $k = 2j+n+1$, respectively. Then we can write the polynomial as

$$\sum_{k\equiv n\,(2)} \binom{m}{k} \left[x + \frac{k-n}{2}\right]_{m-n} \left[x - \frac{k-n}{2}\right]_n + \sum_{k\not\equiv n\,(2)} \binom{m}{k} \left[x - \frac{1}{2} + \frac{k-n}{2}\right]_{m-n} \left[x - \frac{1}{2} - \frac{k-n}{2}\right]_n.$$

Now, let $x = \frac{y}{2}$ for y an odd integer $-1 \le y \le m-1$. Then we get

$$\sum_{k\equiv n\,(2)} \binom{m}{k} \left[\frac{y-n}{2} + \frac{k}{2}\right]_{m-n} \left[\frac{y+n}{2} - \frac{k}{2}\right]_n + \sum_{k\not\equiv n\,(2)} \binom{m}{k} \left[\frac{y-n-1}{2} + \frac{k}{2}\right]_{m-n} \left[\frac{y+n-1}{2} - \frac{k}{2}\right]_n. \tag{14.43}$$

The missing terms for the rest of the k's are all zeros, provided y is an integer of the prescribed size. Hence, we can omit the restrictions on the summations and consider the function

$$P(y,n,m) = \sum_{k=0}^{m} \binom{m}{k} \left[\frac{y-n}{2} + \frac{k}{2}\right]_{m-n} \left[\frac{y+n}{2} - \frac{k}{2}\right]_n, \tag{14.44}$$

with which definition the polynomial in (14.43) for $0 \le y \le m-1$ must coincide with the sum of consecutive values (for y) of P, i.e.,

$$P(y,n,m) + P(y-1,n,m).$$

To evaluate (14.44), we rewrite the factorials using (2.2) twice:

$$\begin{aligned}
\left[\frac{y-n}{2}+\frac{k}{2}\right]_{m-n}&\left[\frac{y+n}{2}-\frac{k}{2}\right]_{n}\\
&=\left[\frac{y-n}{2}+\frac{k}{2}\right]_{k}\left[\frac{y-n}{2}-\frac{k}{2}\right]_{m-n-k}\\
&\quad\times\left[\frac{y+n}{2}-\frac{k}{2}\right]_{m-k}\left[\frac{y+n}{2}+\frac{k}{2}-m\right]_{n-m+k}\\
&=\left[\frac{y-n}{2}+\frac{k}{2}\right]_{k}\left[\frac{y+n}{2}-\frac{k}{2}\right]_{m-k}.
\end{aligned} \tag{14.45}$$

Then we can write (14.44) as

$$P(y,n,m)=\sum_{k=0}^{m}\binom{m}{k}\left[\frac{y-n}{2}+\frac{k}{2}\right]_{k}\left[\frac{y+n}{2}-\frac{k}{2}\right]_{m-k}. \tag{14.46}$$

In the form (14.46) the polynomial is suitable for applying Jensen's formula (14.6). Doing this we get

$$P(y,n,m)=\sum_{j=0}^{m}[m]_j\left(\tfrac{1}{2}\right)^j[y-j]_{m-j}.$$

We apply (2.1), to write

$$[y-j]_{m-j}=(-1)^{m+j}[m-y-1]_{m-j}.$$

Then multiply and divide by $[m-j]_{m-j}$ to get

$$[m-y-1]_{m-j}=[m-j]_{m-j}\binom{m-y-1}{m-j}.$$

Now apply (2.2) and (2.3) to get, for $-1\le y\le m-1$,

$$\begin{aligned}
P(y,n,m)&=(-1)^m\sum_{j=0}^{m}[m]_j[m-j]_{m-j}\binom{m-y-1}{m-j}\left(-\tfrac{1}{2}\right)^j\\
&=(-1)^m[m]_m\left(-\tfrac{1}{2}\right)^{y+1}\sum_{j=0}^{m}\binom{m-y-1}{j-y-1}\left(-\tfrac{1}{2}\right)^{j-y-1}\\
&=(-1)^m[m]_m\left(-\tfrac{1}{2}\right)^{y+1}\left(1-\tfrac{1}{2}\right)^{m-y-1}\\
&=(-1)^{y+m+1}[m]_m\left(\tfrac{1}{2}\right)^m,
\end{aligned} \tag{14.47}$$

where we have applied the binomial formula (7.1). From (14.47) it follows immediately that we have

$$P(y,n,m)+P(y-1,n,m)=0.$$

Hence the polynomial (14.30) has the same values as the polynomial (14.31) in the $m+1$ points, $-1,0,\frac{1}{2},1,\ldots,\frac{m-1}{2}$, so that the polynomials must be identical.

This completes the proof that the polynomial (14.34) is in fact equal to (14.33) independent of n.

Remark 14.8. As soon as this fact is known, it is possible to prove the conjecture by evaluating the polynomial for $n=0$, in which case (14.34) with $m=2\ell$ reduces to

$$\begin{aligned}&\sum_{i=0}^{\ell}\sum_{k=0}^{0}\binom{2\ell+1}{2i-k+1}(x-i)[x+i-k]_{2\ell-0}[x-i+k-1]_{-1}\\&=\sum_{i=0}^{\ell}\binom{2\ell+1}{2i+1}[x-i]_1[x+i]_{2\ell}[x-i-1]_{-1}=\sum_{i=0}^{\ell}\binom{2\ell+1}{2i+1}[x+i]_{2\ell},\end{aligned}\tag{14.48}$$

where we have used (2.2). Now we apply (2.2), (2.3), and (2.5), to write, using (14.30) with $d=2$,

$$\begin{aligned}\binom{2\ell+1}{2i+1}&=\frac{[2\ell+1]_{2i+1}}{[2i+1]_{2i+1}}=\frac{[2\ell+1,2]_{i+1}[2\ell,2]_i}{[2i+1,2]_{i+1}[2i,2]_i}=\frac{\left[\ell+\frac{1}{2}\right]_{i+1}[\ell]_i}{\left[i+\frac{1}{2}\right]_{i+1}[i]_i}\\&=\binom{\ell}{i}\frac{\left[\ell+\frac{1}{2}\right]_{i+1}\left[\ell+\frac{1}{2}\right]_{\ell+1}}{\left[i+\frac{1}{2}\right]_{i+1}\left[\ell+\frac{1}{2}\right]_{\ell+1}}=\frac{1}{\left[\ell-\frac{1}{2}\right]_\ell}\binom{\ell}{i}\left[\ell-\tfrac{1}{2}\right]_i\left[\ell+\tfrac{1}{2}\right]_{\ell-i},\end{aligned}\tag{14.49}$$

and (2.1) and (2.2) to write

$$[x+i]_{2\ell}=[x+i]_i[x]_{2\ell-i}=[-x-1]_i(-1)^i[x]_\ell[x-\ell]_{\ell-i}.\tag{14.50}$$

Hence, we can use (14.49) and (14.50) to write (14.48) as

$$\frac{[x]_\ell}{\left[\ell-\frac{1}{2}\right]_\ell}\sum_{i=0}^{\ell}\binom{\ell}{i}\left[\ell-\tfrac{1}{2}\right]_i[-x-1]_i\left[\ell+\tfrac{1}{2}\right]_{\ell-i}[x-\ell]_{\ell-i}(-1)^i.\tag{14.51}$$

The sum(14.51) satisfies the Pfaff–Saalschütz condition

$$\ell-\tfrac{1}{2}-x-1+\ell+\tfrac{1}{2}+x-\ell-\ell+1=0,$$

so we can apply the Pfaff–Saalschütz identity, cf. (9.1), to obtain

$$\frac{[x]_\ell}{\left[\ell-\frac{1}{2}\right]_\ell}[2\ell]_\ell\left[x-\tfrac{1}{2}\right]_\ell=\frac{[2x,2]_\ell[2x-1,2]_\ell[2\ell]_\ell[\ell]_\ell}{[2\ell-1,2]_\ell[2\ell,2]_\ell}=\frac{[2x]_{2\ell}[2\ell]_{2\ell}}{[2\ell]_{2\ell}}=[2x]_{2\ell}, \tag{14.52}$$

as observed by Joseph Sinyor and Ted Speevak [Sinyor and Speevak 01].

Remark 14.9. This form of the Pfaff–Saalschütz identity is also interesting in itself. Combining (14.48) with (14.52), we get

$$\sum_{i=0}^{\ell}\binom{2\ell+1}{2i+1}[x+i]_{2\ell}=[2x]_{2\ell}. \tag{14.53}$$

This raises two questions about generalizations. First, is the number 2 crucial, or is there a formula

$$\sum_{i=0}^{\left[\frac{m}{2}\right]}\binom{m+1}{2i+1}[x+i]_m=[2x]_m\ ? \tag{14.54}$$

Second, what about the formula

$$\sum_{i=0}^{\left[\frac{m+1}{2}\right]}\binom{m+1}{2i}[x+i]_m=[2x+1]_m\ ? \tag{14.55}$$

They both prove valid. The formulas (14.54) and (14.55) can be reduced to the Pfaff–Saalschütz identity similarly to the proof of (14.53).

Remark 14.10. In [Sinyor and Speevak 01] the authors give a beautiful combinatorial proof of the formula for $n=1$. With the help of Remark 14.9, we can compute this result algebraically from (14.34) with $m=2\ell$. We get

$$\begin{aligned}&\sum_{i=0}^{\ell}\sum_{k=0}^{1}\binom{2\ell}{2i-k+1}(x-i)[x+i-k]_{2\ell-1}[x-i+k-1]_0\\&\quad=\sum_{i=0}^{\ell-1}\binom{2\ell}{2i+1}(x-i)[x+i]_{2\ell-1}+\sum_{i=0}^{\ell}\binom{2\ell}{2i}(x-i)[x+i-1]_{2\ell-1}.\end{aligned} \tag{14.56}$$

We can write $x-i=(2x-2\ell+1)-(x-2\ell+1-i)$ in the first sum and $x-i=2x-(x-i)$ in the second to rewrite the sums in (14.56) conveniently as

$$\begin{aligned}
&\sum_{i=0}^{\ell-1}\binom{2\ell}{2i+1}((2x-2\ell+1)-(x-2\ell+1+i))[x+i]_{2\ell-1}\\
&\qquad+\sum_{i=0}^{\ell}\binom{2\ell}{2i}(2x-(x+i))[x+i-1]_{2\ell-1}\\
&\quad=(2x-2\ell+1)\sum_{i=0}^{\ell-1}\binom{2\ell}{2i+1}[x+i]_{2\ell-1}\\
&\qquad-\sum_{i=0}^{\ell-1}\binom{2\ell}{2i+1}(x-2\ell+1+i)[x+i]_{2\ell-1}\\
&\qquad+2x\sum_{i=0}^{\ell}\binom{2\ell}{2i}[x+i-1]_{2\ell-1}-\sum_{i=0}^{\ell}\binom{2\ell}{2i}(x+i)[x+i-1]_{2\ell-1}\\
&\quad=(2x-2\ell+1)[2x]_{2\ell-1}-\sum_{i=0}^{\ell-1}\binom{2\ell}{2i+1}[x+i]_{2\ell}\\
&\qquad+2x[2(x-1)+1]_{2\ell-1}-\sum_{i=0}^{\ell}\binom{2\ell}{2i}[x+i]_{2\ell}\\
&\quad=[2x]_{2\ell}-\sum_{i=0}^{\ell}\binom{2\ell+1}{2i+1}[x+i]_{2\ell}+[2x]_{2\ell}=[2x]_{2\ell},
\end{aligned}$$

where we have combined the second and fourth sums by (2.4), (2.2), and eventually (14.54) and (14.55).

Remark 14.11. The cases of $n=2\ell$ and $n=2\ell+1$ as considered by Sinyor and Speevak [Sinyor and Speevak 01] are easy to handle from (14.34) with $m=2\ell$. For $n=2\ell$, we obtain exactly the same sum as in (14.56):

$$\begin{aligned}
&\sum_i\sum_{k=2i}^{2i+1}\binom{1}{2i-k+1}\binom{2\ell}{k}(x-i)[x+i-k]_0[x-i+k-1]_{2\ell-1}\\
&\quad=\sum_i\binom{2\ell}{2i}(x-i)[x-i+2i-1]_{2\ell-1}\\
&\qquad\qquad+\sum_i\binom{2\ell}{2i+1}(x-i)[x-i+2i]_{2\ell-1}\\
&\quad=\sum_i\binom{2\ell}{2i}(x-i)[x+i-1]_{2\ell-1}+\sum_i\binom{2\ell}{2i+1}(x-i)[x+i]_{2\ell-1}.
\end{aligned}$$

And for $n = 2\ell + 1$ it is even easier because, as $2i - k + 1 = 0$, the sum (14.34) with $m = 2\ell$ reduces to

$$\sum_i \binom{2\ell+1}{2i+1}[x-i]_1[x-i-1]_{-1}[x+i]_{2\ell} = \sum_i \binom{2\ell+1}{2i+1}[x+i]_{2\ell} = [2x]_{2\ell},$$

using (14.54).

Remark 14.12. We also get an interesting interpolation formula. If we apply (14.45) directly to the first sum in (14.38) with $m = 2\ell > 0$, then we can write it as

$$\sum_j \binom{2\ell}{2j+n}[x+j]_{2j+n}[x-j]_{2\ell-n-2j},$$

and we know its value for $x = -\frac{1}{2}, 0, \frac{1}{2}, 1, \ldots, \ell - \frac{1}{2}$.

If we divide by $(2\ell)!$ and multiply by $2^{2\ell}$, we can write this result for $0 \le n \le 2\ell + 1$ as

$$2^{2\ell} \sum_j \binom{x+j}{2j+n}\binom{x-j}{2\ell-n-2j} = \begin{cases} 0 & \text{for } x = 0, \ldots, \ell, \\ 1 & \text{for } x = -\frac{1}{2}, \ldots, \ell - \frac{1}{2}. \end{cases} \tag{14.57}$$

From (14.39) we know that this polynomial takes the value $2^{2\ell}$ for $x = -1$.

This sum cannot be reduced to the Pfaff–Saalschütz formula.

Remark 14.13. For comparison, the Lagrange interpolation formula gives the polynomial in the form

$$\binom{2x+1}{2\ell+1}(2\ell+1)\sum_{j=0}^{\ell}\binom{2\ell}{2j}\frac{1}{2x+1-2j}. \tag{14.58}$$

The sum in (14.58) can be considered as a generalized Pfaff–Saalschütz sum by the changes

$$\begin{aligned}\binom{2\ell}{2j} &= \frac{[2\ell]_{2j}}{[2j]_{2j}} = \frac{[2\ell,2]_j[2\ell-1,2]_j}{[2j,2]_j[2j-1,2]_j} = \frac{[\ell]_j\left[\ell-\frac{1}{2}\right]_j}{[j]_j\left[j-\frac{1}{2}\right]_j} \\ &= \binom{\ell}{j}\frac{\left[\ell-\frac{1}{2}\right]_j\left[\ell-\frac{1}{2}\right]_\ell}{\left[j-\frac{1}{2}\right]_j\left[\ell-\frac{1}{2}\right]_\ell} = \frac{1}{\left[\ell-\frac{1}{2}\right]_\ell}\binom{\ell}{j}\left[\ell-\frac{1}{2}\right]_j\left[\ell-\frac{1}{2}\right]_{\ell-j},\end{aligned}$$

similar to the changes in (14.49), and

$$\begin{aligned}\frac{1}{2x+1-2j} &= \tfrac{1}{2}\left[x-\tfrac{1}{2}-j\right]_{-1} = -\tfrac{1}{2}\left[-x+\tfrac{1}{2}+j-2\right]_{-1} \\ &= -\tfrac{1}{2}\left[-x-\tfrac{3}{2}+j\right]_j\left[-x-\tfrac{3}{2}\right]_{-j-1} \\ &= -\tfrac{(-1)^j}{2}\left[x+\tfrac{3}{2}-j+j-1\right]_j\left[-x-\tfrac{3}{2}\right]_{-\ell-1} \\ &\quad\times\left[-x-\frac{3}{2}+\ell+1\right]_{\ell-j} \\ &= \frac{(-1)^{j+1}}{2\left[-x-\frac{3}{2}+\ell+1\right]_{\ell+1}}\left[x+\tfrac{1}{2}\right]_j\left[\ell-x-\tfrac{1}{2}\right]_{\ell-j},\end{aligned}$$

where we have applied (2.1) and (2.2).

Then the sum (14.58) becomes

$$\begin{aligned}\frac{-\binom{2x+1}{2\ell+1}(2\ell+1)}{2\left[\ell-\frac{1}{2}\right]_\ell\left[\ell-\frac{1}{2}-x\right]_{\ell+1}}\sum_{j=0}^{\ell}\binom{\ell}{j}\left[\ell-\tfrac{1}{2}\right]_j\left[x+\tfrac{1}{2}\right]_j\left[\ell-\tfrac{1}{2}\right]_{\ell-j} \\ \times\left[\ell-\tfrac{1}{2}-x\right]_{\ell-j}(-1)^j.\end{aligned} \tag{14.59}$$

The factor in (14.59) can be rewritten as

$$\begin{aligned}&-\frac{[2x+1]_{2\ell+1}(2\ell+1)}{[2\ell+1]_{2\ell+1}\,2\left[\ell-\frac{1}{2}\right]_\ell\left[\ell-\frac{1}{2}-x\right]_{\ell+1}} \\ &\qquad= -\frac{[2x+1,2]_{\ell+1}[2x,2]_\ell\,2^{2\ell}[\ell]_\ell}{[2\ell]_{2\ell}\,[2\ell-1,2]_\ell\,[2\ell-1-2x,2]_{\ell+1}\,[\ell]_\ell}, \\ &-\frac{[2x+1,2]_{\ell+1}[x]_\ell\,2^{4\ell}[\ell]_\ell}{[2\ell]_{2\ell}\,[2\ell-1,2]_\ell\,[2\ell,2]_\ell(-1)^{\ell+1}\,[1+2x,2]_{\ell+1}} \\ &\qquad= \frac{(-16)^\ell[x]_\ell}{[2\ell]_\ell[2\ell]_{2\ell}} = \frac{16^\ell[\ell-1-x]_\ell}{[2\ell]_\ell[2\ell]_{2\ell}}.\end{aligned} \tag{14.60}$$

The sum in (14.59) with the factor replaced by (14.60) can be rewritten as a generalized Pfaff–Saalschütz sum, cf. (9.11):

$$\frac{16^\ell[\ell-1-x]_\ell}{[2\ell]_\ell[2\ell]_{2\ell}}\sum_{j=0}^{2\ell}\binom{2\ell}{j}[\ell]_j\left[x+\tfrac{1}{2}\right]_j[-1]_{\ell-j}[-1-x]_{\ell-j}(-1)^j. \tag{14.61}$$

This rewriting may not seem much of an improvement; nevertheless it is possible to simplify it. First denote that $[\ell]_j = 0$ for $j > \ell$, so we can change the upper limit of summation to ℓ. Next, two of the factors in the sum combine as long as $j \leq \ell$:

$$[\ell]_j[-1]_{\ell-j} = [\ell]_j[\ell-j]_{\ell-j}(-1)^{\ell-j} = [\ell]_\ell(-1)^{\ell-j}, \tag{14.62}$$

and two other factors also combine:

$$[\ell-1-x]_\ell[-1-x]_{\ell-j}=[\ell-1-x]_{2\ell-j}. \tag{14.63}$$

Substitution of (14.62) and (14.63) in (14.61) yields

$$\begin{aligned}\frac{(-16)^\ell}{[2\ell]_{2\ell}\binom{2\ell}{\ell}}\sum_{j=0}^{\ell}\binom{2\ell}{j}\left[x+\tfrac{1}{2}\right]_j[\ell-1-x]_{2\ell-j}\\ =\frac{(-16)^\ell}{\binom{2\ell}{\ell}}\sum_{j=0}^{\ell}\binom{x+\frac{1}{2}}{j}\binom{\ell-1-x}{2\ell-j}.\end{aligned} \tag{14.64}$$

In the case of Lagrange interpolation, the Pfaff–Saalschütz formula proved useful to change the formula to some extent. The result is of Chu–Vandermonde type, but the limits of summation are not natural. We could say it is a half Chu–Vandermonde.

The coincidence between the two expressions of this polynomial, (14.57) and (14.64), is a surprising identity.

Chapter 15

Sums of Type V, Harmonic Sums

Harmonic sums are sums whose terms include a generalized harmonic factor, cf. (1.12), i.e., a factor of the form

$$H_{c,n}^{(m)} := \sum_{k=1}^{n} \frac{1}{(c+k)^m}.$$

Note that we consider the harmonic numbers well defined for $n = 0$ as the empty sum $H_{c,0}^{(m)} = 0$.

15.1 Harmonic Sums of Power $m = 1$

Only the simplest harmonic sums allow $c \neq 0$. The first indefinite harmonic sum is the sum of harmonic numbers with factorial factors.

Theorem 15.1. *For all $m, n \in \mathbb{Z}$, $m \neq -1$,*

$$\sum [c+n+k]_m H_{c,n+k}^{(1)} \delta k = \frac{[c+n+k]_{m+1}}{m+1} \left(H_{c,n+k}^{(1)} - \frac{1}{m+1} \right). \quad (15.1)$$

Proof: Taking the difference of the right side, we find the proof is straightforward. □

The omitted case is covered by the following result.

Theorem 15.2. *For all $n \in \mathbb{Z}$,*

$$\sum [c+n+k]_{-1} H_{c,n+k}^{(1)} \delta k = \tfrac{1}{2} \left(\left(H_{c,n+k}^{(1)} \right)^2 - H_{c,n+k}^{(2)} \right). \quad (15.2)$$

Proof: Taking the difference of the right side, the proof is again straightforward. □

Corollary 15.3. *For* $c = 0$,

$$\sum_{k=1}^{n} \frac{H_k}{k} = \tfrac{1}{2}\left(H_n^2 + H_{0,n}^{(2)}\right).$$

This is the form preferred by D. E. Knuth, R. L. Graham, and O. Patashnik [Graham et al. 94, (6.71)].

Remark 15.4. The formulas (15.1) and (15.2) allow the evaluation of sums of the form

$$\sum \frac{p(k)}{q(k)} H_{c,k}^{(1)} \delta k,$$

with a rational factor, provided the denominator has different roots. We just apply (5.4).

Theorem 15.5. *For* $c \in \mathbb{C}$ *and* $n \in \mathbb{Z}$, *we have the indefinite summation formula*

$$\sum \left(H_{c,n+k}^{(1)}\right)^2 \delta k = (c+n+k)\left(H_{c,n+k}^{(1)}\right)^2 - (2(c+n+k)+1)H_{c,n+k}^{(1)} + 2(c+n+k). \tag{15.3}$$

Proof: Straightforward. □

Theorem 15.6. *For all* $x \in \mathbb{C} \setminus \{0\}$,

$$\sum (-1)^{k-1}\binom{x}{k} H_k \delta k = (-1)^k \left(\binom{x-1}{k-1} H_k - \frac{1}{x}\binom{x-1}{k}\right). \tag{15.4}$$

Proof: Straightforward. □

This is actually a special case of a more general formula:

Theorem 15.7. *For all* $x, y \in \mathbb{C}$, $y \notin \mathbb{N}$ *such that* $x - y - 1 \neq 0$,

$$\sum \frac{[x]_k}{[y]_k} H_{-y-1,k}^{(1)} \delta k = \frac{1}{x-y-1} \frac{[x]_k}{[y]_{k-1}} \left(H_{-y-1,k-1}^{(1)} + \frac{1}{x-y-1}\right). \tag{15.5}$$

Proof: Straightforward. □

For $x \in \mathbb{N}$, we can generalize the harmonic factor:

Theorem 15.8. *For all $n \in \mathbb{N}$, $m \in \mathbb{Z}$, and $c \in \mathbb{C}$,*

$$\sum_{k=0}^{n} (-1)^{k-1} \binom{n}{k} H^{(1)}_{c,m+k} = \frac{(n-1)!}{[c+m+n]_n} = \frac{1}{n\binom{c+m+n}{n}}. \tag{15.6}$$

Proof: Let $m = 0$. We apply the formula (2.14) and summation by parts (2.17) to write the sum as

$$-\sum_{k=0}^{n} (-1)^{k-1} \binom{n-1}{k} \frac{1}{c+k+1} = \sum_{k=0}^{n} (-1)^{k} \binom{n-1}{k} [c+k]_{-1}.$$

This is a Chu–Vandermonde sum (8.11), so it equals

$$\frac{(n-1)!}{[c+n]_n}.$$

□

The special case of $c = 0$ is found in [Kaucký 75, (6.7.1)].

Inversion of formula (15.6) (cf. (2.18), (2.19)) for $m = 0$ yields a formula of type II(2, 2, 1) which we could have known already:

Corollary 15.9. *For all $n \in \mathbb{N}$ and $c \in \mathbb{C}$,*

$$\sum_{k=1}^{n} \frac{1}{k} \cdot \frac{[n]_k}{[-c-1]_k} = H^{(1)}_{c,n}.$$

Proof: Let us define

$$f(n) = \sum_{k=1}^{n} \frac{1}{k} \cdot \frac{[n]_k}{[-c-1]_k}.$$

Then we observe that the upper limit is natural, so the difference is

$$\begin{aligned} f(n) - f(n-1) &= \sum_{k=1}^{n} \frac{[n-1]_{k-1}}{[-c-1]_k} \\ &= \frac{1}{n} \cdot \frac{1}{c+n} \left[\frac{[n]_k}{[-c-1]_{k-1}} \right]_1^{n+1} \\ &= \frac{1}{n} \cdot \frac{1}{c+n}(0-n) = \frac{1}{c+n}, \end{aligned}$$

where we have used formula (7.3).

□

Theorem 15.10. *We have the following indefinite summation formula for any $x \in \mathbb{C}$, but in case $x \in \mathbb{N}$, we require $k \leq n$:*

$$\sum(-1)^k \binom{x}{k}^{-1} H_k \delta k = \frac{(-1)^k k}{x+2} \binom{x}{k-1}^{-1} \left(\frac{1}{x+2} - H_k\right). \tag{15.7}$$

Proof: Straightforward. □

The following formulas are found in [Kaucký 75, (6.7.6), (6.7.7)]. They are very special cases of Theorem 15.8 for the choice of $x = 2n$ and $x = 2n - 1$, respectively, and limits 1 and $2n - 1$.

Theorem 15.11. *For any integer $n \in \mathbb{N}$,*

$$\sum_{k=1}^{2n}(-1)^{k-1}\binom{2n}{k}^{-1} H_k = \frac{n}{2(n+1)^2} + \frac{1}{2n+2} H_{2n},$$

$$\sum_{k=1}^{2n-1}(-1)^{k-1}\binom{2n-1}{k}^{-1} H_k = \frac{2n}{2n+1} H_{2n}.$$

Theorem 15.10 can be generalized to:

Theorem 15.12. *For all $x, y \in \mathbb{C}$ such that $x - y - 1 \neq 0$, $x \notin \mathbb{N}$,*

$$\sum \frac{[x]_k}{[y]_k} H^{(1)}_{-x-1,k} \delta k = \frac{1}{x-y-1} \frac{[x]_k}{[y]_{k-1}} \left(H^{(1)}_{-x-1,k} + \frac{1}{x-y-1}\right). \tag{15.8}$$

Proof: Straightforward. □

The following formula is found in [Kaucký 75, (6.7.3)].

Theorem 15.13. *For all $n \in \mathbb{N}$,*

$$\sum_{k=0}^{n}(-1)^{k-1}\binom{n}{k} H_{2k} = \frac{1}{2n} + \frac{[2n-2,2]_{n-1}}{[2n-1,2]_n}.$$

Proof: We have from (2.13)

$$\Delta(-1)^k \binom{n-1}{k-1} = (-1)^{k-1}\binom{n}{k}.$$

Summation by parts, therefore, yields

$$-\sum_{k=0}^{n}(-1)^{k+1}\binom{n-1}{k}\left(\frac{1}{2k+2} + \frac{1}{2k+1}\right)$$

$$= \frac{1}{2}\sum_{k=0}^{n-1}(-1)^k \binom{n-1}{k}[k]_{-1} + \frac{1}{2}\sum_{k=0}^{n-1}(-1)^k\binom{n-1}{k}\left[-\tfrac{1}{2}+k\right]_{-1}$$

$$= \frac{(-1)^{n-1}}{2}\left([0]_{-n}[-1]_{n-1} + \left[-\tfrac{1}{2}\right]_{-n}[-1]_{n-1}\right) = \frac{1}{2n} + \frac{[n-1]_{n-1}}{2\left[n-\frac{1}{2}\right]_n},$$

where we have applied the Chu–Vandermonde convolution, (8.11). The formula follows after multiplication by the appropriate power of 2. □

The following is a generalization of a strange formula due to W. Ljunggren from 1947 [Ljunggren 47] which was only proved in 1948 by J. E. Fjeldstad [Fjeldstad 48] and J. Kvamsdal [Kvamsdal 48] (cf. [Kaucký 75, (6.7.5)]), using analytic methods.

Theorem 15.14. *For any complex number $x \in \mathbb{C}$ and any integers $n \in \mathbb{N}$ and $p \in \mathbb{N}_0$,*

$$\sum_{k=p}^{n} \binom{n-p}{k-p}\binom{x+p}{k} H_k = \sum_{k=0}^{n-p} \binom{n-p}{k}\binom{x+p}{k+p} H_{k+p}$$
$$= \binom{x+n}{n}\left(H_n + H_{x,p}^{(1)} - H_{x,n}^{(1)}\right).$$

From this theorem Ljunggren's formula follows with the choices $x = n$ and $p = 0$:

Corollary 15.15. *For any integer, $n \in \mathbb{N}$,*

$$\sum_{k=1}^{n} \binom{n}{k}^2 H_k = \binom{2n}{n}\left(H_n - H_{n,n}^{(1)}\right) = \binom{2n}{n}(2H_n - H_{2n}).$$

Proof of Theorem 15.14: We consider the function

$$f(n,x,p) = \sum_{k=p}^{n} \binom{n-p}{k-p}\binom{x+p}{k} H_k, \tag{15.9}$$

which we shall compute. Now we remark that

$$\Delta_k (-1)^k \binom{x}{k} H_k = (-1)^{k+1}\binom{x+1}{k+1}\left(H_{k+1} - \frac{1}{x+1}\right). \tag{15.10}$$

Using (15.10), summation by parts (2.17) yields

$$f(n,x,p) = -\sum_{k=p}^{n} (-1)^k \binom{n-p-1}{k-p}(-1)^{k+1}\binom{x+p+1}{k+1}$$
$$\times\left(H_{k+1} - \frac{1}{x+p+1}\right)$$
$$= \sum_{k=p+1}^{n-1} \binom{n-p-1}{k-p-1}\binom{x+p+1}{k} H_k - \frac{1}{x+p+1}$$
$$\times \sum_{k=p}^{n-1} \binom{n-p-1}{n-1-k}\binom{x+p+1}{k+1}.$$

The first sum is just $f(n,x,p+1)$ as defined in (15.9), while the second is a Chu–Vandermonde sum as in (8.1), so it yields

$$\frac{1}{x+p+1}\binom{x+n}{n}.$$

Hence, we have derived the formula with the difference taken in the variable p,

$$\Delta_p f(n,x,p) = \frac{1}{x+p+1}\binom{x+n}{n}.$$

Summation gives us

$$\begin{aligned} f(n,x,p) &= f(n,x,n) - \binom{x+n}{n}\sum_{k=p}^{n-1}\frac{1}{x+k+1} \\ &= \binom{x+n}{n}H_n - \binom{x+n}{n}H^{(1)}_{x+p,n-p} \\ &= \binom{x+n}{n}\left(H_n + H^{(1)}_{x,p} - H^{(1)}_{x,n}\right), \end{aligned}$$

proving the formula. □

A more general result is the following formula:

Theorem 15.16. *For any complex number $x \in \mathbb{C}$,*

$$\sum_{k=1}^{n}\binom{n}{k}[-x]_k[x]_{n-k}H_k = (-1)^n(n-1)! - \frac{1}{n}[x-1]_n.$$

Proof: From Corollary 6.2 we have the difference formula

$$\binom{n}{k}[-x]_k[x]_{n-k} = \Delta\binom{n-1}{k-1}[-x-1]_{k-1}[x]_{n-k+1}. \tag{15.11}$$

Summation by parts (2.17) now yields

$$\begin{aligned} -[x]_n - \sum_{k=1}^{n}\binom{n-1}{k}[-x-1]_k[x]_{n-k}\frac{1}{k+1} \\ = -[x]_n + \tfrac{1}{n}\sum_{k=1}^{n}\binom{n}{k+1}[-x]_{k+1}[x-1]_{n-k-1}. \end{aligned}$$

This is a Chu–Vandermonde sum, missing the first two terms corresponding to $k=0$ and $k=-1$. Hence, we get

$$-[x]_n + \frac{1}{n}\left([-1]_n - n(-x)[x-1]_{n-1} - [x-1]_n\right) = (-1)^n(n-1)! - \tfrac{1}{n}[x-1]_n.$$

This ends the proof. □

Remark 15.17. If $x \in \mathbb{N}_0$, $x \leq n$, the second term vanishes, so the right side becomes just

$$(-1)^n(n-1)!,$$

independent of x. The case $x = n$ is found in [Kaucký 75, (6.7.2)].

A similar but more specialized formula is the following:

Theorem 15.18. *For all $p \in \mathbb{N}$,*

$$\sum_{k=0}^{n} \binom{n}{k} [-p]_k [p]_{n-k} H_{p+k-1} = (-1)^n (n-1)!. \tag{15.12}$$

Proof: We write the sum as

$$[p]_n H_{p-1} + \sum_{k=1}^{n} \binom{n}{k} [-p]_k [p]_{n-k} H_{p+k-1}.$$

We have from (15.11),

$$\binom{n}{k} [-p]_k [p]_{n-k} = \Delta \binom{n-1}{k-1} [-p-1]_{k-1} [p]_{n-k+1}.$$

Hence, summation by parts yields

$$\begin{aligned}
[p]_n H_{p-1} - [p]_n H_p &- \sum_{k=1}^{n} \binom{n-1}{k} [-p-1]_k [p]_{n-k} \cdot \frac{1}{p+k} \\
&= -[p-1]_{n-1} - \sum_{k=1}^{n} \binom{n-1}{k} [-p-1]_{k-1} [p]_{n-k} \\
&= -[p-1]_{n-1} + \sum_{k=1}^{n} \binom{n-1}{k} [-p]_k [p-1]_{n-1-k} \\
&= -[p-1]_{n-1} + [-1]_{n-1} - [p-1]_{n-1} = (-1)^n (n-1)!,
\end{aligned}$$

where we have moved a factor p and applied the Chu–Vandermonde convolution, (8.3). □

The special case of $p = n$ is similar to Remark 15.17 above.

Remark 15.19. For all $n \in \mathbb{N}$,

$$\sum_{k=0}^{n} \binom{n}{k} \binom{-n}{k} H_{n+k-1} = \frac{(-1)^n}{n}.$$

Proof: Chose $p = n$ in (15.12) and divide by $n!$. □

A similar formula not containing any of the preceding three as special cases is found in [Gould 72b, (7.15)].

Theorem 15.20. *For $x, y \in \mathbb{C}$ where x and $x + y$ are not negative integers,*

$$\sum_{k=0}^{n} \binom{n}{k} [x]_k [y+n]_{n-k} H_{y,k}^{(1)} = [x+y+n]_n \left(H_{y,n}^{(1)} - H_{x+y,n}^{(1)} \right). \quad (15.13)$$

Proof: We proceed by induction on n.

For $n = 1$ the result is obvious. Denote the left-hand side of (15.13) by S:

$$S(n,x,y) = \sum_{k=0}^{n} \binom{n}{k} [x]_k [y+n]_{n-k} H_{y,k}^{(1)}.$$

Splitting the binomial coefficient using (2.7), we get S as a sum of two terms:

$$(1) = \sum_{k=0}^{n-1} \binom{n-1}{k} [x]_k [y+n]_{n-k} H_{y,k}^{(1)},$$

$$(2) = \sum_{k=1}^{n} \binom{n-1}{k-1} [x]_k [y+n]_{n-k} H_{y,k}^{(1)}.$$

The first is

$$\begin{aligned}(1) &= (y+n) S(n-1,x,y) \\ &= (y+n)[x+y+n-1]_{n-1} \left(H_{y,n-1}^{(1)} - H_{x+y,n-1}^{(1)} \right).\end{aligned}$$

In the second we shift the index by 1 to get

$$(2) = \sum_{k=0}^{n-1} \binom{n-1}{k} [x]_{k+1} [y+1+n-1]_{n-1-k} H_{y,k+1}^{(1)}.$$

We remark that $H^{(1)}_{y,k+1} = \frac{1}{y+1} + H^{(1)}_{y+1,k}$, so we can write

$$\begin{aligned}
(2) &= x\sum_{k=0}^{n-1}\binom{n-1}{k}[x-1]_k[y+1+n-1]_{n-1-k}H^{(1)}_{y+1,k} \\
&\quad + x\sum_{k=0}^{n-1}\binom{n-1}{k}[x-1]_k[y+1+n-1]_{n-1-k}\frac{1}{y+1} \\
&= xS(n-1,x-1,y+1) + \frac{x}{y+1}[x+y+n-1]_{n-1} \\
&= x[x+y+n-1]_{n-1}\left(H^{(1)}_{y+1,n-1} - H^{(1)}_{x+y,n-1} + \frac{1}{y+1}\right) \\
&= x[x+y+n-1]_{n-1}\left(H^{(1)}_{y,n-1} - H^{(1)}_{x+y,n-1} + \frac{1}{y+n}\right),
\end{aligned}$$

applying (15.13) to the first sum and the Chu–Vandermonde convolution (8.3) to the second sum.

Adding (1) and (2) we get

$$\begin{aligned}
(x+y+n)&[x+y+n-1]_{n-1}\left(H^{(1)}_{y,n-1} - H^{(1)}_{x+y,n-1}\right) \\
&\quad + \frac{x}{y+n}[x+y+n-1]_{n-1} \\
&=[x+y+n]_n\left(H^{(1)}_{y,n-1} - H^{(1)}_{x+y,n-1}\right) + [x+y+n]_n \\
&\quad \times\left(\frac{1}{y+n} - \frac{1}{x+y+n}\right) \\
&=[x+y+n]_n\left(H^{(1)}_{y,n} - H^{(1)}_{x+y,n}\right).
\end{aligned}$$

□

Remark 15.21. The analog of the sum of harmonic numbers with alternating signs does not provide anything new. Actually we have the formula,

$$\sum_{k=1}^{n}\frac{(-1)^k}{k} = H_{\lfloor\frac{n}{2}\rfloor} - H_n.$$

15.2 Harmonic Sums of Power $m > 1$

There are a few known formulas containing the generalized harmonic numbers (1.12) of power $m = 2$. Their proofs are straightforward.

Theorem 15.22. *For all $x \in \mathbb{C}$,*

$$\sum \binom{x}{k}\binom{-x}{k} H_{0,k}^{(2)} \delta k = \binom{x-1}{k-1}\binom{-x-1}{k-1} H_{0,k}^{(2)} + \frac{1}{x^2}\binom{x-1}{k}\binom{-x-1}{k}.$$

Corollary 15.23. *For all $n \in \mathbb{N}$,*

$$\sum_{k=0}^{n} \binom{n}{k}\binom{-n}{k} H_{0,k}^{(2)} = -\frac{1}{n^2}.$$

Sums containing a factorial with nonnegative index can also be found.

Theorem 15.24. *For $m \in \mathbb{N}_0$, we have the indefinite sum*

$$\sum [c+k]_m H_{c,k}^{(2)} \delta k = \frac{[c+k]_{m+1}}{m+1} H_{c,k}^{(2)} - \frac{(-1)^m m!}{m+1} H_{c,k}^{(1)} - \sum_{j=1}^{m} \frac{(-1)^j [m]_{m-j} [c+k]_j}{j(m+1)}. \tag{15.14}$$

15.3 Bang Seung-Jin's Problem

Bang Seung-Jin [Bang 95] posed the following problem in 1995.

Show that

$$\sum_{k=1}^{n} \frac{(-1)^{k-1}}{k}\binom{n}{k} \sum_{j=1}^{k} \frac{1}{j}\left(\frac{1}{1}+\frac{1}{2}+\cdots+\frac{1}{j}\right) = \sum_{k=1}^{n} \frac{1}{k^3},$$

for all positive integers n.

This is the special case $p = 2$ of the following:

Theorem 15.25. *For $n \in \mathbb{N}$ and $p \in \mathbb{N}_0$, define the multiple sum*

$$S(n,p) = \sum_{k=1}^{n} \frac{(-1)^{k-1}}{k}\binom{n}{k} \sum_{j_1=1}^{k} \frac{1}{j_1} \sum_{j_2=1}^{j_1} \frac{1}{j_2} \sum_{j_3=1}^{j_2} \cdots \frac{1}{j_{p-1}} \sum_{j_p=1}^{j_{p-1}} \frac{1}{j_p}.$$

Then we have

$$S(n,p) = \sum_{k=1}^{n} \frac{1}{k^{p+1}} = H_{0,n}^{(p+1)}.$$

Proof: We consider the difference $D(n,p) = S(n,p) - S(n-1,p)$,

$$D(n,p) = \sum_{k=1}^{n} \frac{(-1)^{k-1}}{k}\left(\binom{n}{k} - \binom{n-1}{k}\right) \sum_{j_1=1}^{k} \frac{1}{j_1} \sum_{j_2=1}^{j_1} \frac{1}{j_2} \sum_{j_3=1}^{j_2} \cdots \sum_{j_p=1}^{j_{p-1}} \frac{1}{j_p},$$

and we want to prove that it satisfies

$$D(n,p) = \frac{1}{n^{p+1}}.$$

Since

$$\frac{1}{k}\left(\binom{n}{k} - \binom{n-1}{k}\right) = \frac{1}{k}\binom{n-1}{k-1} = \frac{1}{n}\binom{n}{k},$$

the difference can be written (where for convenience we have added the zero term for $k=0$)

$$D(n,p) = \frac{1}{n}\sum_{k=0}^{n}(-1)^{k-1}\binom{n}{k} \sum_{j_1=1}^{k} \frac{1}{j_1} \sum_{j_2=1}^{j_1} \frac{1}{j_2} \sum_{j_3=1}^{j_2} \cdots \frac{1}{j_{p-1}} \sum_{j_p=1}^{j_{p-1}} \frac{1}{j_p}.$$

As the constant terms vanish, summation by parts or Abelian summation (2.17) yields

$$\begin{aligned} D(n,p) &= -\frac{1}{n}\sum_{k=0}^{n}(-1)^{k+1}\binom{n-1}{k}\frac{1}{k+1} \sum_{j_2=1}^{k+1} \frac{1}{j_2} \sum_{j_3=1}^{j_2} \cdots \frac{1}{j_{p-1}} \sum_{j_p=1}^{j_{p-1}} \frac{1}{j_p} \\ &= \frac{1}{n^2}\sum_{k=0}^{n-1}(-1)^{k}\binom{n}{k+1} \sum_{j_2=1}^{k+1} \frac{1}{j_2} \sum_{j_3=1}^{j_2} \cdots \frac{1}{j_{p-1}} \sum_{j_p=1}^{j_{p-1}} \frac{1}{j_p} \\ &= \frac{1}{n^2}\sum_{k=0}^{n}(-1)^{k-1}\binom{n}{k} \sum_{j_2=1}^{k} \frac{1}{j_2} \sum_{j_3=1}^{j_2} \cdots \frac{1}{j_{p-1}} \sum_{j_p=1}^{j_{p-1}} \frac{1}{j_p} = \frac{1}{n}D(n,p-1), \end{aligned}$$

i.e., the difference satisfies the recursion

$$D(n,p) = \frac{1}{n}D(n,p-1).$$

Hence we get

$$D(n,p) = \frac{1}{n^p}D(n,0) = \frac{1}{n^{p+1}}\sum_{k=1}^{n}(-1)^{k-1}\binom{n}{k},$$

where we must change the limit to 1, because the zero term no longer vanishes.

But from (2.13) we have

$$\sum_{k=1}^{n}(-1)^{k-1}\binom{n}{k} = (-1)^k\binom{n-1}{k-1}\Bigg]_1^{n+1} = 0 - (-1) = 1.$$

Alternatively, it follows from the binomial theorem (7.1) as

$$\sum_{k=0}^{n}(-1)^k\binom{n}{k} = (1-1)^n = 0. \tag{15.15}$$

This proves the theorem. □

15.4 The Larcombe Identities

In [Larcombe et al. 05, Larcombe and Larsen 07], we prove a family of harmonic identities.

For non-negative integers n and p and a complex number z, not equal to $-1, -2, \ldots, -n$, let

$$S(n,p,z) = z\binom{z+n}{n}\sum_{k=0}^{n}(-1)^k\binom{n}{k}\frac{1}{(z+k)^p}.$$

Theorem 15.26.

$$S(n,0,z) = z0^n.$$

Proof: Apply the binomial formula for $(1-1)^n$. □

Theorem 15.27.

$$S(n,1,z) = 1.$$

Proof:

$$\begin{aligned}
S(n,1,z) &= \frac{[z+n]_{n+1}}{n!}\sum_{k=0}^{n}\frac{n!}{k!(n-k)!}(-1)^k\frac{1}{z+k} \\
&= \sum_{k=0}^{n}\frac{1}{k!(n-k)!}(-1)^k[z+n]_{n-k}[z+k-1]_k \\
&= \sum_{k=0}^{n}\binom{-z}{k}\binom{n+z}{n-k} = \binom{n}{n} = 1,
\end{aligned}$$

using Chu–Vandermonde convolution (8.1). □

Theorem 15.28. *For $p > 0$ we have the recursion*

$$S(n,p,z) = \sum_{k=0}^{n}\frac{1}{z+k}S(k,p-1,z).$$

Corollary 15.29. *For $p > 0$ we have the recursion*

$$\Delta_k S(k-1, p, z) = \frac{1}{z+k} S(k, p-1, z).$$

Proof of Theorem 15.28:

$$\begin{aligned} S(n,p,z) &= \frac{[z+n]_{n+1}}{n!} \sum_{k=0}^{n} \binom{n}{k} (-1)^k \frac{1}{(z+k)^p} \\ &= \sum_{k=0}^{n} \binom{-z}{k} \binom{z+n}{n-k} \frac{1}{(z+k)^{p-1}} \\ &= \sum_{k=0}^{n} \binom{-z}{k} \binom{-z-k-1}{n-k} (-1)^{n-k} \frac{1}{(z+k)^{p-1}}. \end{aligned}$$

Now we apply Chu–Vandermonde convolution to write

$$\binom{-z-k-1}{n-k} = \sum_{j=k}^{n} \binom{-z-k}{j-k} \binom{-1}{n-j} = \sum_{j=k}^{n} \binom{-z-k}{j-k} (-1)^{n-j}.$$

Recalling (2.9)

$$\binom{-z}{k} \binom{-z-k}{j-k} = \binom{-z}{j} \binom{j}{k},$$

changing the order of summation, and using (2.11), (2.10), and (15.15), we obtain

$$\begin{aligned} S(n,p,z) &= \sum_{j=0}^{n} \binom{-z}{j} (-1)^j \sum_{k=0}^{j} \binom{j}{k} (-1)^k \frac{1}{(z+k)^{p-1}} \\ &= \sum_{j=0}^{n} \binom{z+j-1}{j} \sum_{k=0}^{j} \binom{j}{k} (-1)^k \frac{1}{(z+k)^{p-1}} \\ &= \sum_{j=0}^{n} \frac{1}{z+j} z \binom{z+j}{j} \sum_{k=0}^{j} \binom{j}{k} (-1)^k \frac{1}{(z+k)^{p-1}} \\ &= \sum_{j=0}^{n} \frac{1}{z+j} S(j, p-1, z). \end{aligned}$$

□

Theorem 15.30. *For $p > 1$*

$$S(n,p,z) = \frac{1}{p-1} \sum_{j=1}^{p-1} H_{z,n}^{(j)} S(n, p-j, z).$$

Proof: We proceed by induction on p.

For $p = 2$ we have

$$S(n,2,z) = \sum_{k=0}^{n} \frac{1}{z+k} S(k,1,z) = \sum_{k=0}^{n} \frac{1}{z+k} = H_{z,n}^{(1)}$$
$$= \frac{1}{2-1} \sum_{j=1}^{2-1} H_{z,n}^{(j)} S(n, 2-j, z)$$

by Theorem 15.27, then Theorem 15.26, and eventually the definition of harmonic numbers (1.12). The induction step goes

$$\begin{aligned}
&\Delta_k \frac{1}{p-1} \sum_{j=1}^{p-1} H_{z,k-1}^{(j)} S(k-1, p-j, z) \\
&= \frac{1}{p-1} \sum_{j=1}^{p-1} \frac{1}{(z+k)^j} S(k, p-j, z) \\
&\quad + \frac{1}{p-1} \sum_{j=1}^{p-2} H_{z,k}^{(j)} \frac{1}{z+k} S(k, p-1-j, z) \\
&\quad - \frac{1}{p-1} \sum_{j=1}^{p-2} \frac{1}{(z+k)^{j+1}} S(k, p-1-j, z) \\
&= \frac{1}{p-1} \frac{1}{z+k} S(k, p-1, z) + \frac{1}{p-1} \sum_{j=2}^{p-1} \frac{1}{(z+k)^j} S(k, p-j, z) \\
&\quad + \frac{1}{p-1} (p-2) \frac{1}{z+k} S(k, p-1, z) \\
&\quad - \frac{1}{p-1} \sum_{j=2}^{p-1} \frac{1}{(z+k)^j} S(k, p-j, z) \\
&= \frac{1}{z+k} S(k, p-1, z).
\end{aligned}$$

Use the formula (2.15) on each term of the sum and Corollary 15.29 for ΔS. Take the first term outside of the first sum. Apply the induction assumption for $p-1$ on the second sum. Change the summation index by 1 in the last sum. Then add the first and third expressions with the second and last sums cancelling. □

Corollary 15.31.

$$\begin{aligned}
S(n,1,z) =&1,\\
S(n,2,z) =&H^{(1)}_{z,n},\\
S(n,3,z) =&\frac{1}{2}\left(\left(H^{(1)}_{z,n}\right)^2 + H^{(2)}_{z,n}\right),\\
S(n,4,z) =&\frac{1}{6}\left(\left(H^{(1)}_{z,n}\right)^3 + 3H^{(1)}_{z,n}H^{(2)}_{z,n} + 2H^{(3)}_{z,n}\right),\\
S(n,5,z) =&\frac{1}{24}\Bigg(\left(H^{(1)}_{z,n}\right)^4 + 6\left(H^{(1)}_{z,n}\right)^2 H^{(2)}_{z,n} + 3\left(H^{(2)}_{z,n}\right)^2 + 8H^{(1)}_{z,n}H^{(3)}_{z,n}\\
&+ 6H^{(4)}_{z,n}\Bigg),\\
S(n,6,z) =&\frac{1}{120}\Bigg(\left(H^{(1)}_{z,n}\right)^5 + 10\left(H^{(1)}_{z,n}\right)^3 H^{(2)}_{z,n} + 15H^{(1)}_{z,n}\left(H^{(2)}_{z,n}\right)^2\\
&+20\left(H^{(1)}_{z,n}\right)^2 H^{(3)}_{z,n} + 20H^{(2)}_{z,n}H^{(3)}_{z,n} + 30H^{(1)}_{z,n}H^{(4)}_{z,n} + 24H^{(5)}_{z,n}\Bigg),\\
S(n,7,z) =&\frac{1}{720}\Bigg(\left(H^{(1)}_{z,n}\right)^6 + 15\left(H^{(1)}_{z,n}\right)^4 H^{(2)}_{z,n} + 45\left(H^{(1)}_{z,n}\right)^2\left(H^{(2)}_{z,n}\right)^2\\
&+ 40\left(H^{(1)}_{z,n}\right)^3 H^{(3)}_{z,n} + 120H^{(1)}_{z,n}H^{(2)}_{z,n}H^{(3)}_{z,n} + 90\left(H^{(1)}_{z,n}\right)^2 H^{(4)}_{z,n}\\
&+ 144H^{(1)}_{z,n}H^{(5)}_{z,n} + 15\left(H^{(2)}_{z,n}\right)^3 + 90H^{(2)}_{z,n}H^{(4)}_{z,n} + 40\left(H^{(3)}_{z,n}\right)^2\\
&+ 120H^{(6)}_{z,n}\Bigg).
\end{aligned}$$

Proof: Let us do $S(n,5,z)$. Take

$$H^{(1)}_{z,n}S(n,4,z)/4 = \frac{1}{24}\left(\left(H^{(1)}_{z,n}\right)^4 + 3\left(H^{(1)}_{z,n}\right)^2 H^{(2)}_{z,n} + 2H^{(1)}_{z,n}H^{(3)}_{z,n}\right).$$

Then take

$$\begin{aligned}
H^{(2)}_{z,n}S(n,3,z)/4 &= \frac{1}{8}\left(\left(H^{(1)}_{z,n}\right)^2 H^{(2)}_{z,n} + \left(H^{(2)}_{z,n}\right)^2\right)\\
&= \frac{1}{24}\left(3\left(H^{(1)}_{z,n}\right)^2 H^{(2)}_{z,n} + 3\left(H^{(2)}_{z,n}\right)^2\right).
\end{aligned}$$

So take

$$H^{(3)}_{z,n}S(n,2,z)/4 = \frac{1}{24}6H^{(1)}_{z,n}H^{(3)}_{z,n},$$

and eventually

$$H^{(4)}_{z,n}S(n,1,z)/4 = \frac{1}{24}6H^{(4)}_{z,n}.$$

Adding, we get the coefficients 1, $3+3=6$, 3, $2+6=8$, and 6. □

Theorem 15.32. *The coefficients in the corollary are*

$$\frac{(p-1)!}{\prod_i i^{n_i} n_i!} \prod_i \left(H_{z,n}^{(i)}\right)^{n_i},$$

with $\sum_i i n_i = p-1$.

Proof: Guess the coefficients as class orders from symmetric groups, cf. [Stanley 97]. □

Appendix A

Indefinite Sums

A.1 Rational Functions

In principle, rational functions can always be summed indefinitely. The function can be split into a polynomial part and a sum of principal fractions. All we need are the formulas (3.4) ff., (1.12), and (3.32):

$$\sum [c+k]_m \delta k = \begin{cases} \frac{[c+k]_{m+1}}{m+1} & m \neq -1, \\ H_{c,k}^{(1)} & m = -1, \end{cases}$$

$$\sum \frac{1}{(c+k)^m} \delta k = H_{c,k}^{(m)},$$

$$\sum k^m \delta k = \frac{B_{m+1}(k)}{m+1}.$$

Rational functions may take the convenient form, cf. (7.3),

$$\sum \frac{[a]_k}{[b]_k} \delta k = \frac{1}{a-b-1} \frac{[a]_k}{[b]_{k-1}},$$

to be recognized as a sum of type I with quotient

$$q(k) = \frac{a-k}{b-k}.$$

A more complicated form with a condition on the arguments is the following, cf. (8.27). If $p = a+b-c-d \in \mathbb{N}_0$, we have

$$\sum \frac{[a]_k [b]_k}{[c-1]_k [d-1]_k} \delta k = \frac{[a]_k [b]_k}{[c-1]_{k-1} [d-1]_{k-1}} \sum_{j=0}^{p} \frac{[p]_j [k-c-1]_j}{[a-c]_{j+1} [b-c]_{j+1}},$$

with quotient

$$q(k) = \frac{(a-k)(b-k)}{(c-1-k)(d-1-k)}.$$

A similar, but more special, formula is

$$\sum \binom{x}{k}(x-2k)\delta k = k\binom{x}{k},$$

with quotient

$$q(k) = \frac{(x-k)\left(\frac{x}{2}-1-k\right)}{(-1-k)\left(\frac{x}{2}-k\right)}(-1).$$

Furthermore, we have the surprising formula, cf. (11.1):

$$\sum \frac{[a+b+c]_k[a]_k[b]_k[c]_k(a+b+c-2k)(-1)^k}{k![b+c-1]_k[a+c-1]_k[a+b-1]_k}\delta k$$
$$= \frac{[a+b+c]_k[a-1]_{k-1}[b-1]_{k-1}[c-1]_{k-1}(-1)^k}{k![b+c-1]_{k-1}[a+c-1]_{k-1}[a+b-1]_{k-1}}.$$

A.2 Other Indefinite Sums

We know a few indefinite sums involving harmonic numbers (cf. (15.1), (15.2), (15.3)–(15.5), (15.7), (15.8), and (15.14)). Seven involve first-order harmonic numbers:

$$\sum [c+k]_m H_{c,k}^{(1)}\delta k = \frac{[c+k]_{m+1}}{m+1}\left(H_{c,k}^{(1)} - \frac{1}{m+1}\right),$$

$$\sum [c+k]_{-1} H_{c,k}^{(1)}\delta k = \tfrac{1}{2}\left(\left(H_{c,k}^{(1)}\right)^2 - H_{c,k}^{(2)}\right),$$

$$\sum \left(H_{c,k}^{(1)}\right)^2\delta k = (c+k)\left(H_{c,k}^{(1)}\right)^2 - (2(c+k)+1)H_{c,k}^{(1)} + 2(c+k),$$

$$\sum (-1)^{k-1}\binom{x}{k}H_k\delta k = (-1)^k\left(\binom{x-1}{k-1}H_k - \frac{1}{x}\binom{x-1}{k}\right),$$

$$\sum \frac{[x]_k}{[y]_k}H_{-y-1,k}^{(1)}\delta k = \frac{1}{x-y-1}\frac{[x]_k}{[y]_{k-1}}\left(H_{-y-1,k-1}^{(1)} + \frac{1}{x-y-1}\right),$$

$$\sum \frac{[x]_k}{[y]_k}H_{-x-1,k}^{(1)}\delta k = \frac{1}{x-y-1}\frac{[x]_k}{[y]_{k-1}}\left(H_{-x-1,k}^{(1)} + \frac{1}{x-y-1}\right),$$

$$\sum (-1)^k\binom{x}{k}^{-1}H_k\delta k = \frac{(-1)^k k}{x+2}\binom{x}{k-1}^{-1}\left(\frac{1}{x+2} - H_k\right).$$

Three involve second-order harmonic numbers:

$$\sum H_{c,k}^{(2)}\delta k = (c+k)H_{c,k}^{(2)} - H_{c,k}^{(1)},$$
$$\sum (c+k)H_{c,k}^{(2)}\delta k = \frac{1}{2}\left([c+k]_2 H_{c,k}^{(2)} + H_{c,k}^{(1)} - (c+k)\right),$$
$$\sum [c+k]_2 H_{c,k}^{(2)}\delta k = \frac{1}{3}\left([c+k]_3 H_{c,k}^{(2)} - 2H_{c,k}^{(1)} + 2(c+k) - \frac{[c+k]_2}{2}\right).$$

The general form for $m \in \mathbb{N}_0$ is

$$\sum [c+k]_m H_{c,k}^{(2)}\delta k = \frac{[c+k]_{m+1}}{m+1} H_{c,k}^{(2)} - \frac{(-1)^m m!}{m+1} H_{c,k}^{(1)} - \sum_{j=1}^{m} \frac{(-1)^j [m]_{m-j}[c+k]_j}{j(m+1)}.$$

Appendix B

Basic Identities

In this appendix, we will present basic generalizations of the simplest ordinary combinatorial identities. We prove the generalizations from which the ordinary cases are deduced simply by substitution of $Q = R = 1$ in the generalizations.

Our definition of a *basic number* is slightly more general than the usual definition. The main reason for the generalization is that it makes many formulas more symmetric in their appearance.

We will assume the existence of two "universal constants", Q and R. We define the *basic transformation* $\{x\}$ of x by

$$\{x\} = \begin{cases} xQ^{x-1} & \text{if } Q = R, \\ (Q^x - R^x)/(Q - R) & \text{if } Q \neq R. \end{cases} \tag{B.1}$$

The following properties of $\{x\}$ are easily verified:

$$\{0\} = 0 \quad \text{and} \quad \{1\} = 1\,, \tag{B.2}$$

$$\{-x\} = -\{x\}(QR)^{-x}\,, \tag{B.3}$$

$$\{x + y\} = \{x\}Q^y + \{y\}R^x = \{x\}R^y + \{y\}Q^x\,, \tag{B.4}$$

$$\begin{aligned} \{x - y\} &= \{x\}R^{-y} - \{y\}R^{-y}Q^{x-y} \\ &= \{x\}Q^{-y} - \{y\}Q^{-y}R^{x-y}\,. \end{aligned} \tag{B.5}$$

If n is a positive integer then

$$\begin{aligned} \{n\} &= Q^{n-1} + Q^{n-2}R + \cdots + R^{n-1}, \\ \{-n\} &= -(Q^{-1}R^{-n} + Q^{-2}R^{-n+1} + \cdots + Q^{-n}R^{-1}). \end{aligned} \tag{B.6}$$

Remark B.1. Because of the symmetry in Q and R in the definition (B.1), Q and R must be interchangeable in all the following formulas.

Remark B.2. The expression (B.6) shows that our definition generalizes the definition of $[k]_{p,q}$ due to Wachs and White [Wachs and White 91].

In his original introduction to *basic numbers* in 1847 [Heine 47], E. Heine replaced the parameters α, β, γ in a Gaussian hypergeometric series

$$1+\frac{(1-q^{\alpha})(1-q^{\beta})}{(1-q)(1-q^{\gamma})}x+\frac{(1-q^{\alpha})(1-q^{\alpha+1})(1-q^{\beta})(1-q^{\beta+1})}{(1-q)(1-q^{2})(1-q^{\gamma})(1-q^{\gamma+1})}x^{2}+\cdots,$$

by

$$\frac{1-q^{\alpha}}{1-q},\quad \frac{1-q^{\beta}}{1-q},\quad \frac{1-q^{\gamma}}{1-q},$$

respectively, in order to obtain a continuous generalization from α, as

$$\frac{1-q^{\alpha}}{1-q}\to\alpha,$$

etc., for $q\to 1$. He also noticed the similarity between q and $r=\frac{1}{q}$.

We have replaced q by Q/R to obtain

$$\frac{1-q^{\alpha}}{1-q}=\frac{R^{\alpha}-Q^{\alpha}}{R-Q}R^{1-\alpha}=\{\alpha\}R^{1-\alpha},$$

hardly a generalization at all. Nevertheless, some of our formulas include two of Heine's by choosing $R=q$ and $Q=1$ or $R=1$ and $Q=q$.

It is sometimes customary to omit the denominator $1-q$, because it often cancels out anyway. We have chosen not to do so to keep the continuity. It is further customary to replace q^{α} with a, and further $q^{\alpha+1}$ with aq, etc., but we will not do so, because as Heine pointed out in 1878, it makes no difference [Heine 78, p. 98]. One just has to apply the usual formulas to q^{α} for comparison.

B.1 Factorials and Binomial Coefficients

We now introduce a *generalized basic factorial*,

$$\{x,d\}_n=\begin{cases}\prod_{j=0}^{n-1}\{x-jd\} & \text{if } n\in\mathbb{N},\\ 1 & \text{if } n=0,\\ \prod_{j=1}^{-n}\frac{1}{\{x+jd\}} & \text{if } -n\in\mathbb{N}.\end{cases}$$

Since descending factorials are the most important in combinatorics, we will use the notation $\{x\}_n$ for $\{x,1\}_n$. If $Q=R=1$, we get the ordinary descending factorial $[x]_n$. The following are the most important properties of the generalized basic factorial:

$$\{x,d\}_n = \{x-d,d\}_n Q^{nd} + \{x-d,d\}_{n-1}\{nd\}R^{x-nd}, \tag{B.7}$$

$$\{x,d\}_n = \{x-d,d\}_n Q^{nd} - \{x-d,d\}_{n-1}\{-nd\}R^{x}Q^{nd}, \tag{B.8}$$

$$\{x,d\}_n = \{-x+(n-1)d,d\}_n(-1)^n(QR)^{nx-\binom{n}{2}d}, \tag{B.9}$$

$$\{x,d\}_n = \{x-(n-1)d,-d\}_n, \tag{B.10}$$

$$\{x,d\}_n = \{-x,-d\}_n(-1)^n(QR)^{nx-\binom{n}{2}d}, \tag{B.11}$$

$$\{x,d\}_{m+n} = \{x,d\}_m\{x-md,d\}_n, \tag{B.12}$$

$$\{x,d\}_{-n} = 1/\{x+nd,d\}_n, \tag{B.13}$$

$$\frac{\{2x,2d\}_n}{\{x,d\}_n} = \prod_{j=0}^{n-1}(Q^{x-jd}+R^{x-jd}), \tag{B.14}$$

$$\begin{aligned}\{2nd,d\}_n &= \{2nd-d,2d\}_n\frac{\{(2nd,2d\}_n}{\{nd,d\}_n}\\ &= \{-d,2d\}_n(-1)^n(QR)^{n^2d}\prod_{j=1}^{n}(Q^{jd}+R^{jd}).\end{aligned} \tag{B.15}$$

The *basic binomial coefficients* are defined by

$$\left\{{x \atop n}\right\} = \begin{cases}\{x\}_n/\{n\}_n & \text{if } n\in\mathbb{N}_0,\\ 0 & \text{if } n\notin\mathbb{N}_0.\end{cases} \tag{B.16}$$

We note the following properties:

$$\left\{{n \atop 0}\right\} = \left\{{n \atop n}\right\} = 1, \tag{B.17}$$

$$\left\{{x \atop n}\right\} = \left\{{x-1 \atop n-1}\right\}Q^{x-n} + \left\{{x-1 \atop n}\right\}R^n, \tag{B.18}$$

$$\left\{{x \atop n}\right\}\left\{{n \atop m}\right\} = \left\{{x \atop m}\right\}\left\{{x-m \atop n-m}\right\}, \tag{B.19}$$

$$\left\{{x \atop n}\right\}\{n\}_m = \{x\}_m\left\{{x-m \atop n-m}\right\}. \tag{B.20}$$

If $x\in\mathbb{N}_0$ and $x\geq n$, then $\left\{{x \atop n}\right\} = \left\{{x \atop x-n}\right\}$, and from (B.18) it follows that $\left\{{x \atop n}\right\}$ is a homogeneous polynomial in Q and R of degree $(x-n)n$.

B.2 Two Basic Binomial Theorems

Theorem B.3. *For arbitrary x and y, integers m and n ($n \geq m$) and arbitrary α and β,*

$$\sum_{k=m}^{n} \left\{ {n-m \atop k-m} \right\} x^{k-m} y^{n-k} Q^{\binom{k-m}{2}} R^{\binom{n-k}{2}} = \prod_{j=0}^{n-m-1} (xQ^j + yR^j), \tag{B.21}$$

$$\sum_{k=m}^{n} \left\{ {n-m \atop k-m} \right\} x^{k-m} y^{n-k} Q^{\binom{k-m}{2}+\alpha(k-m)} R^{\binom{n-k}{2}+\beta(n-k)} = \prod_{j=0}^{n-m-1} (xQ^{j+\alpha} + yR^{j+\beta}). \tag{B.22}$$

Proof: It is sufficient to prove (B.21) for $m = 0$, because the general case follows by substitution of $k + m$ for k and $n + m$ for n. Let

$$S(n, x, y) = \sum_{k} \left\{ {n \atop k} \right\} x^k y^{n-k} Q^{\binom{k}{2}} R^{\binom{n-k}{2}}.$$

Using

$$\left\{ {n \atop k} \right\} = \left\{ {n-1 \atop k-1} \right\} Q^{n-k} + \left\{ {n-1 \atop k} \right\} R^k,$$

we split S into two sums. In the sum with $\left\{ {n-1 \atop k-1} \right\}$ we can substitute $k + 1$ for k. When we take advantage of common factors in the two sums, this gives us

$$\begin{aligned} S(n, x, y) &= \sum_{k} \left\{ {n-1 \atop k} \right\} x^k y^{n-1-k} Q^{\binom{k}{2}} R^{\binom{n-k}{2}} (xQ^{n-1} + yR^{n-1}) \\ &= S(n-1, x, y)(xQ^{n-1} + yR^{n-1}). \end{aligned}$$

Iteration now yields the formula (B.21).

The more general formula (B.22) follows from (B.21) by substitution of xQ^α for x and yR^β for y. □

B.3 Two Basic Chu–Vandermonde Convolutions

Theorem B.4. *For arbitrary a and b and arbitrary integers $0 \le m \le n$,*

$$\sum_{k=m}^{n} \left\{ {n-m \atop k-m} \right\} \{a\}_{k-m} \{b\}_{n-k} (QR)^{-(k-m)(n-k)} Q^{a(n-k)} R^{b(k-m)} = \{a+b\}_{n-m}, \tag{B.23}$$

$$\sum_{k=m}^{n} \left\{ {n-m \atop k-m} \right\} \{a,-1\}_{k-m} \{b,-1\}_{n-k} Q^{a(n-k)} R^{b(k-m)} = \{a+b,-1\}_{n-m}. \tag{B.24}$$

Proof: It is sufficient to prove (B.23) and (B.24) for $m = 0$. We will prove (B.24), from which (B.23) follows.

Let

$$S(n,a,b) = \sum_{k} \left\{ {n \atop k} \right\} \{a,-1\}_k \{b,-1\}_{n-k} Q^{a(n-k)} R^{bk}.$$

Using (B.18),

$$\left\{ {n \atop k} \right\} = \left\{ {n-1 \atop k-1} \right\} R^{n-k} + \left\{ {n-1 \atop k} \right\} Q^k,$$

we split S into two sums. In the sum with $\left\{ {n-1 \atop k-1} \right\}$ we substitute $k+1$ for k. When we take advantage of common factors in the two sums, this gives us

$$S(n,a,b) = \sum_{k} \left\{ {n-1 \atop k} \right\} \{a,-1\}_k \{b,-1\}_{n-1-k} Q^{a(n-1-k)} R^{bk} \times (\{a+k\} R^{b+n-1-k} + \{b+n-1-k\} Q^{a+k}).$$

Using (B.4) this gives

$$S(n,a,b) = S(n-1,a,b)\{a+b+n-1\}).$$

Iteration now yields formula (B.24).

If we apply (B.24) to $-a$ and $-b$ using (B.6) to change the sign of d, we get (B.23). □

B.4 Two Special Cases of the Basic Chu–Vandermonde Convolutions

Theorem B.5. *For arbitrary c, an arbitrary integer p, and integers m and n with $n \geq m$,*

$$\sum_{k=m}^{n} \begin{Bmatrix} n-m \\ k-m \end{Bmatrix} (-1)^{n-k} \{c-m+k\}_p Q^{(n-m)(p-c-n+m)+\binom{n-k}{2}} \times R^{(n-k)(p+m-n+1)+\binom{n-k}{2}} = \{p\}_{n-m}\{c\}_{p-n+m}, \qquad \text{(B.25)}$$

$$\sum_{k=m}^{n} \begin{Bmatrix} n-m \\ k-m \end{Bmatrix} (-1)^{k-m} \{c+n-k\}_p Q^{(n-m)(p-c-n+m)+\binom{k-m}{2}} \times R^{(k-m)(p+m-n+1)+\binom{k-m}{2}} = \{p\}_{n-m}\{c\}_{p-n+m}. \qquad \text{(B.26)}$$

Proof: In (B.25) we substitute $-1-c$ for a and $c-m+n-p$ for b, and using (B.9) on the right-hand side we obtain:

$$\sum_{k} \begin{Bmatrix} n-m \\ k-m \end{Bmatrix} \{-1-c\}_{k-m}\{c-m+n-p\}_{n-k} \times Q^{(-1-c-k+m)(n-k)} R^{(c-m-p+k)(k-m)} = \{n-m-p-1\}_{n-m}$$
$$= \{p\}_{n-m}(-1)^{n-m}(QR)^{(n-m-p-1)(n-m)-\binom{n-m}{2}}.$$

In the sum we use formula (B.9) on the factor $\{-1-c\}_{k-m}$ and observe that

$$\{c-m+k\}_{k-m}\{c-m+n-p\}_{n-k} = \{c-m+k\}_p\{c-m+n-p\}_{n-m-p},$$

since all factors occur with the same multiplicity on both sides. We obtain

$$\sum_{k} \begin{Bmatrix} n-m \\ k-m \end{Bmatrix} (-1)^{k-m}\{c-m+k\}_p\{c-m+n-p\}_{n-m-p} \times Q^{(-1-c-k+m)(n-k)} R^{(c-m-p+k)(k-m)} (QR)^{(-1-c)(k-m)-\binom{k-m}{2}}$$
$$= \{p\}_{n-m}(-1)^{n-m}(QR)^{(n-m-p-1)(n-m)-\binom{n-m}{2}}. \qquad \text{(B.27)}$$

Now multiply (B.27) by $\{c\}_{p-n+m} = 1/\{c-m+n-p\}_{n-m-p}$ and by $(-1)^{n-m}(QR)^{-(n-m-p-1)(n-m)+\binom{n-m}{2}}$. Simplifying the exponents of Q and R yields (B.25).

The equation (B.26) is obtained from (B.25) by substitution of $n+m-k$ for k, i.e., by reversal of the direction of summation. □

Formula (B.25) for $p = -1$ has a nice interpretation as a generalization of a useful partition of the quotient into partial fractions.

Corollary B.6. *For arbitrary c and arbitrary nonnegative integer n,*

$$\frac{1}{\{c+n\}_{n+1}} = \frac{1}{\{n\}_n} \sum_{k=0}^{n} \left\{ n \atop k \right\} \frac{(-1)^k}{\{c+k\}} (QR)^{\binom{k+1}{2}} Q^{-n(c+k)}. \tag{B.28}$$

Proof: Substitution of $m = 0$, $p = -1$, and $c = c - 1$ gives the formula. □

B.5 The Symmetric Kummer Identity

Theorem B.7. *For arbitrary a and arbitrary nonnegative integer n,*

$$\sum_{k=0}^{2n} \left\{ 2n \atop k \right\} \{a\}_k \{a\}_{2n-k} (-1)^k (QR)^{-k(2n-k)} = \{2a, 2\}_n \{-1, 2\}_n \tag{B.29}$$

$$= (-1)^n \{2a, 2\}_n \{2n-1, 2\}_n (QR)^{-n^2},$$

$$\sum_{k=0}^{2n+1} \left\{ 2n+1 \atop k \right\} \{a\}_k \{a\}_{2n+1-k} (-1)^k (QR)^{-k(2n+1-k)} = 0. \tag{B.30}$$

Proof: Let

$$S(n, a) = \sum_k \left\{ n \atop k \right\} \{a\}_k \{a\}_{n-k} (-1)^k (QR)^{-k(n-k)}.$$

Then observe that

$$S(2n+1, a) = \sum_k \left\{ 2n+1 \atop k \right\} \{a\}_k \{a\}_{2n+1-k} (-1)^k (QR)^{-k(2n+1-k)}$$

$$= -S(2n+1, a) = 0,$$

by reversing the order of summation.

Now, by (B.18).

$$S(2n, a) = \sum_k \left\{ 2n-1 \atop k \right\} \{a\}_k \{a\}_{2n-k} (-1)^k Q^{-k(2n-1-k)} R^{-k(2n-k)}$$

$$+ \sum_k \left\{ 2n-1 \atop k-1 \right\} \{a\}_k \{a\}_{2n-k} (-1)^k Q^{-k(2n-k)} R^{-(k-1)(2n-k)}.$$

Replacing k with $k+1$, the second sum becomes

$$-\sum_k \left\{ {2n-1 \atop k} \right\} \{a\}_{k+1} \{a\}_{2n-1-k} (-1)^k Q^{-(k+1)(2n-1-k)} R^{-k(2n-1-k)}.$$

Application of (B.12) with $m=1$ to $\{a\}_{k+1}$ and $\{a\}_{2n-k}$ yields

$$\begin{aligned} S(2n,a) &= \{a\} \sum_k \left\{ {2n-1 \atop k} \right\} \{a\}_k \{a-1\}_{2n-1-k} (-1)^k Q^{-k(2n-1-k)} \\ &\quad \times R^{-k(2n-k)} - \{a\} \sum_k \left\{ {2n-1 \atop k} \right\} \{a-1\}_k \{a\}_{2n-1-k} (-1)^k \\ &\quad \times Q^{-(k+1)(2n-1-k)} R^{-k(2n-1-k)}. \end{aligned}$$

We apply (B.7) in the forms

$$\begin{aligned} \{a\}_k &= \{a-1\}_k R^k + \{a-1\}_{k-1} \{k\} Q^{a-k}, \\ \{a\}_{2n-1-k} &= \{a-1\}_{2n-1-k} Q^{2n-1-k} \\ &\qquad + \{a-1\}_{2n-2-k} \{2n-1-k\} R^{a-2n+1+k}, \end{aligned}$$

and get

$$\begin{aligned} &S(2n,a) \\ &= \{a\} \Bigg(\sum_k \left\{ {2n-1 \atop k} \right\} \{a-1\}_k \{a-1\}_{2n-1-k} (-1)^k \\ &\qquad\qquad \times (QR)^{-k(2n-1-k)} \\ &\quad + \sum_k \left\{ {2n-1 \atop k} \right\} \{a-1\}_{k-1} \{a-1\}_{2n-1-k} \{k\} (-1)^k Q^a \\ &\qquad\qquad \times (QR)^{-k(2n-k)} \\ &\quad - \sum_k \left\{ {2n-1 \atop k} \right\} \{a-1\}_k \{a-1\}_{2n-1-k} (-1)^k \\ &\qquad\qquad \times (QR)^{-k(2n-1-k)} \\ &\quad - \sum_k \left\{ {2n-1 \atop k} \right\} \{a-1\}_k \{a-1\}_{2n-2-k} \{2n-1-k\} (-1)^k \\ &\qquad\qquad \times R^a (QR)^{-(k+1)(2n-1-k)} \Bigg). \end{aligned}$$

The first and third sums cancel. To the remaining two sums we apply the formula (B.20) to get

$$\begin{aligned}
&S(2n,a)\\
&\quad= \{a\}\{2n-1\}\\
&\qquad\times\Bigg(\sum_k \left\{ {2n-2 \atop k-1} \right\}\{a-1\}_{k-1}\{a-1\}_{2n-1-k}(-1)^k Q^a (QR)^{-k(2n-k)}\\
&\qquad-\sum_k \left\{ {2n-2 \atop k} \right\}\{a-1\}_k\{a-1\}_{2n-2-k}(-1)^k R^a (QR)^{-(k+1)(2n-1-k)}\Bigg)\\
&\quad= -\{a\}\{2n-1\}\,(Q^a+R^a)\\
&\qquad\quad\times\sum_k \left\{ {2n-2 \atop k} \right\}\{a-1\}_k\{a-1\}_{2n-2-k}(-1)^k\\
&\qquad\quad\times(QR)^{-k(2n-2-k)+1-2n}\\
&\quad= -\{a\}\{2n-1\}\,(Q^a+R^a)\,S(2n-2,a-1)(QR)^{1-2n},
\end{aligned}$$

by replacing k with $k+1$ in the first sum. From the definition (B.1) we have

$$\{a\}\,(Q^a+R^a) = \{2a\}.$$

Because $S(0,x)=1$, iteration of the reduction yields

$$S(2n,a) = (-1)^n\{2a,2\}_n\{2n-1,2\}_n(QR)^{-n^2} = \{2a,2\}_n\{-1,2\}_n.$$

The last equality is due to (B.9). □

The symmetry of the formula (B.29) is emphasized by the following change of summation variable.

Corollary B.8. *For arbitrary a and arbitrary nonnegative integer n,*

$$\begin{aligned}
\sum_{k=-n}^{n} \left\{ {2n \atop n+k} \right\}\{a\}_{n+k}\{a\}_{n-k}(-1)^k(QR)^{k^2} &= \{2a,2\}_n\{2n-1,2\}_n\\
&= (-1)^n\{2a,2\}_n\{-1,2\}_n(QR)^{n^2}.
\end{aligned}$$

B.6 The Quasi-Symmetric Kummer Identity

Theorem B.9. *For arbitrary a, arbitrary integer p and arbitrary nonnegative integers $m \leq n$,*

$$\sum_{k=m}^{n} \begin{Bmatrix} n-m \\ k-m \end{Bmatrix} \{a\}_{k-m}\{a+p\}_{n-k}(-1)^{k-m}(QR)^{-(k-m)(n-k)}R^{(k-m)p}$$
$$= (\sigma(p))^{n-m} \sum_{j=\lceil \frac{n-m-|p|}{2} \rceil}^{\lfloor \frac{n-m}{2} \rfloor} \begin{Bmatrix} |p| \\ n-m-2j \end{Bmatrix}$$
$$\times \{n-m\}_{n-m-2j}\{2(a+(p\wedge 0)),2\}_j\{-1,2\}_j$$
$$\times Q^{(a+(p\wedge 0)-2j)(n-m-2j)}R^{(|p|-n+m+2j)2j+(n-m)(p\wedge 0)}. \tag{B.31}$$

Proof: It is enough to prove the theorem for $m = 0$. We apply the Chu–Vandermonde convolution (B.23) on $\{a+p\}_{n-k}$ in the form

$$\{a+p\}_{n-k} = \sum_j \begin{Bmatrix} n-k \\ j-k \end{Bmatrix} \{a\}_{j-k}\{p\}_{n-j}Q^{(a-j+k)(n-j)}R^{(p-n+j)(j-k)},$$

to get

$$\sum_k \begin{Bmatrix} n \\ k \end{Bmatrix} \{a\}_k(-1)^k \sum_j \begin{Bmatrix} n-k \\ j-k \end{Bmatrix} \{a\}_{j-k}\{p\}_{n-j}$$
$$\times Q^{(a-j+k)(n-j)}R^{(p-n+j)(j-k)}$$
$$= \sum_j \{p\}_{n-j} \sum_k (-1)^k \begin{Bmatrix} n \\ k \end{Bmatrix} \begin{Bmatrix} n-k \\ j-k \end{Bmatrix} \{a\}_k\{a\}_{j-k}$$
$$\times Q^{(a-j+k)(n-j)}R^{(p-n+j)(j-k)}.$$

Next (B.19) allows the transformation

$$\begin{Bmatrix} n \\ k \end{Bmatrix} \begin{Bmatrix} n-k \\ j-k \end{Bmatrix} = \begin{Bmatrix} n \\ n-k \end{Bmatrix} \begin{Bmatrix} n-k \\ n-j \end{Bmatrix} = \begin{Bmatrix} n \\ n-j \end{Bmatrix} \begin{Bmatrix} j \\ j-k \end{Bmatrix} = \begin{Bmatrix} n \\ j \end{Bmatrix} \begin{Bmatrix} j \\ k \end{Bmatrix}.$$

Hence, we obtain from (B.29)

$$\sum_j \{p\}_{n-j} \begin{Bmatrix} n \\ j \end{Bmatrix} Q^{(a-j)(n-j)}R^{(p-n+j)j}$$
$$\times \sum_k (-1)^k \begin{Bmatrix} j \\ k \end{Bmatrix} \{a\}_k\{a\}_{j-k}(QR)^{-k(j-k)}$$
$$= \sum_j \{p\}_{n-2j} \begin{Bmatrix} n \\ 2j \end{Bmatrix} \{2a,2\}_j\{-1,2\}_j Q^{(a-2j)(n-2j)}R^{(p-n+2j)2j}.$$

In the case $p \geq 0$, we can write

$$\{p\}_{n-2j}\left\{ n \atop 2j \right\} = \left\{ p \atop n-2j \right\}\{n\}_{n-2j},$$

and note that

$$\left\{ p \atop n-2j \right\} = 0 \text{ for } j < \frac{n-p}{2}.$$

In the case of negative p, we apply the above formula for $a+p$ and $-p$. □

B.7 The Balanced Kummer Identity

Theorem B.10. *For arbitrary a and arbitrary integer $n \geq 0$,*

$$\sum_{k=0}^{n} \left\{ n \atop k \right\} \{n+a\}_k \{n-a\}_{n-k} (-1)^k (QR)^{k(k-n-a)} = \{n-a-1, 2\}_n \frac{\{2n, 2\}_n}{\{n\}_n}. \tag{B.32}$$

In order to prove this theorem, we need a 2-step version of formula (B.25).

Lemma B.11. *For arbitrary c and arbitrary nonnegative integer n,*

$$\frac{\{2n, 2\}_n}{\{2(c+n), 2\}_{n+1}} = \sum_{k=0}^{n} \frac{\{2n, 2\}_k}{\{2k, 2\}_k} \frac{(-1)^k}{\{2(c+k)\}} (QR)^{k(k+1)} Q^{-2n(c+k)}. \tag{B.33}$$

Proof: We consider the formula (B.28) for the universal constants Q^2 and R^2, in which case we denote the basic transformation by $\{\{\cdot\}\}$, e.g.,

$$\{\{x\}\} = \frac{(Q^{2x} - R^{2x})}{(Q^2 - R^2)} = \frac{\{2x\}}{(Q+R)}. \tag{B.34}$$

Then we have

$$\{\{x\}\}_n = \frac{\{2x, 2\}_n}{(Q+R)^n}, \tag{B.35}$$

and hence,

$$\left\{\left\{ x \atop n \right\}\right\} = \frac{\{\{x\}\}_n}{\{\{n\}\}_n} = \frac{\{2x, 2\}_n}{\{2n, 2\}_n}. \tag{B.36}$$

Therefore, we get

$$\frac{1}{\{\{c+n\}\}_{n+1}} = \frac{1}{\{\{n\}\}_n} \sum_{k=0}^{n} \left\{\left\{ n \atop k \right\}\right\} \frac{(-1)^k}{\{\{c+k\}\}} (QR)^{k(k+1)} Q^{-2n(c+k)}.$$

Substitution of (B.34)–(B.36) gives (B.33). □

Proof of Theorem B.10: Consider the sum

$$S := \sum_k \left\{ {n \atop k} \right\} \{n+a\}_k \{n-a\}_{n-k} (-1)^k (QR)^{k(k-n-a)}.$$

We apply (B.11) and (B.10) to write

$$\begin{aligned} \{n+a\}_k(-1)^k &= \{-a-n,-1\}_k (QR)^{k(n+a)-\binom{k}{2}} \\ &= \{-a-n-1+k\}_k (QR)^{k(n+a)-\binom{k}{2}}. \end{aligned}$$

The sum then looks like

$$S = \sum_k \left\{ {n \atop k} \right\} \{n-a\}_{n-k} \{-a-n-1+k\}_k (QR)^{\binom{k+1}{2}}.$$

Now we apply (B.12) to write

$$\{n-a\}_{2n+1} = \{n-a\}_{n-k} \{-a+k\}_{n+1} \{-a-n-1+k\}_k.$$

The sum is then written

$$S = \{n-a\}_{2n+1} \sum_k \left\{ {n \atop k} \right\} \frac{(QR)^{\binom{k+1}{2}}}{\{-a+k\}_{n+1}}.$$

Now we apply Corollary B.6 to

$$\frac{1}{\{-a+k\}_{n+1}} = \frac{1}{\{n\}_n} \sum_j \left\{ {n \atop j} \right\} \frac{(-1)^j}{\{-a-n+k+j\}} (QR)^{\binom{j+1}{2}} \times Q^{-n(-a-n+k+j)}.$$

Substitution of $j = i-k$ into this sum and this sum into the sum above yields

$$S = \{n-a\}_{2n+1} \sum_k \left\{ {n \atop k} \right\} \frac{1}{\{n\}_n} \sum_i \left\{ {n \atop i-k} \right\} \frac{(-1)^{i-k}}{\{-a-n+i\}} \times (QR)^{\binom{k+1}{2}+\binom{i-k+1}{2}} Q^{-n(-a-n+i)}.$$

Interchanging the order of summation gives

$$S = \frac{\{n-a\}_{2n+1}}{\{n\}_n} \sum_i \frac{(-1)^i}{\{-a-n+i\}} Q^{-n(-a-n+i)} (QR)^{\binom{i+1}{2}} \times \sum_k \left\{ {n \atop k} \right\} \left\{ {n \atop i-k} \right\} (-1)^k (QR)^{k^2-ik}.$$

Using the definition of the binomial coefficients (B.16), we recognize the second sum as an example of the symmetric Kummer expression, (B.29) and (B.30). Hence, the odd sums vanish, and the even sums can be written with $2j$ substituted for i according to (B.29):

$$\begin{aligned}
&\sum_k \left\{ {n \atop k} \right\} \left\{ {n \atop i-k} \right\} (-1)^k (QR)^{k^2-ik} \\
&= \frac{1}{\{i\}_i} \sum_k \left\{ {i \atop k} \right\} \{n\}_k \{n\}_{i-k} (-1)^k (QR)^{-k(i-k)} \\
&= \frac{1}{\{2j\}_{2j}} \sum_k \left\{ {2j \atop k} \right\} \{n\}_k \{n\}_{2j-k} (-1)^k (QR)^{-k(2j-k)} \\
&= \frac{1}{\{2j\}_{2j}} (-1)^j \{2n,2\}_j \{2j-1,2\}_j (QR)^{-j^2} \\
&= \frac{\{2n,2\}_j}{\{2j,2\}_j} (-1)^j (QR)^{-j^2}.
\end{aligned}$$

With this reduction we obtain

$$S = \frac{\{n-a\}_{2n+1}}{\{n\}_n} \sum_j \frac{\{2n,2\}_j}{\{2j,2\}_j} \frac{(-1)^j}{\{-a-n+2j\}} (QR)^{j^2+j} Q^{-n(-a-n+2j)}.$$

To this sum we can apply Lemma B.11 with $c = \frac{-a-n}{2}$. □

B.8 The Quasi-Balanced Kummer Identity

Theorem B.12. *For arbitrary a and arbitrary integers p and $n \geq m \geq 0$, let*

$$\begin{aligned}
S = \sum_{k=m}^{n} \left\{ {n-m \atop k-m} \right\} &\{n-m+a-p\}_{k-m} \{n-m-a\}_{n-k} (-1)^{k-m} \\
&\times (QR)^{(k-m)(k-n-a)} Q^{(k-m)p}.
\end{aligned}$$

Then for $p \geq 0$,

$$\begin{aligned}
S = \frac{\{2n-2m-2p,2\}_{n-m-p}}{\{n-m-p\}_{n-m-p}} &\sum_{j=0}^{p} \left\{ {p \atop j} \right\} \{n-m-a-1+j,2\}_{n-m} \\
&\times Q^{(p-j)(n-m-p)+\binom{p+1-j}{2}} R^{-pj+\binom{j+1}{2}}, \qquad \text{(B.37)}
\end{aligned}$$

and for $p < 0$,

$$S = \frac{\{2n-2m,2\}_{n-m}}{\{n-m-p\}_{n-m-p}} R^{-p(a+n-m)+\binom{1-p}{2}}$$
$$\times \sum_{j=0}^{-p} \left\{ {-p \atop j} \right\} (-1)^j \{n-m-p-a-1-j,2\}_{n-m-p}$$
$$\times\, Q^{j(n-m)+\binom{j+1}{2}} R^{\binom{j}{2}}. \qquad \text{(B.38)}$$

Proof: It is enough to prove the theorem for $m = 0$.

(1) For $p \geq 0$, let

$$S(n,p,a) = \sum_k \left\{ {n \atop k} \right\} \{n+a-p\}_k \{n-a\}_{n-k} (-1)^k Q^{k(k-n-a+p)} R^{k(k-n-a)}.$$

We split this sum in two by applying (4.2) to

$$\{n-a\}_{n-k} = \{n-a-1\}_{n-k} R^{n-k} + \{n-k\}\{n-a-1\}_{n-k-1} Q^{k-a},$$

to get

$$S(n,p,a) = \sum_k \left\{ {n \atop k} \right\} \{n-p+a\}_k \{n-a-1\}_{n-k} (-1)^k$$
$$\times\, Q^{k(k-n+p-a)} R^{k(k-n-a-1)} R^n$$
$$+\{n\} \sum_k \left\{ {n-1 \atop k} \right\} \{n-p+a\}_k \{n-a-1\}_{n-k-1} (-1)^k$$
$$\times\, Q^{k(k-n+p-a+1)-a} R^{k(k-n-a)}.$$

And we split the sum $S(n,p,a+1)$ by applying (B.7) to

$$\{n-p+a+1\}_k = \{n-p+a\}_k Q^k + \{k\}\{n-p+a\}_{k-1} R^{n-p+a+1-k},$$

to get

$$S(n,p,a+1) = \sum_k \left\{ {n \atop k} \right\} \{n-p+a\}_k \{n-a-1\}_{n-k} (-1)^k$$
$$\times\, Q^{k(k-n+p-a)} R^{k(k-n-a-1)}$$
$$+\{n\} \sum_k \left\{ {n-1 \atop k-1} \right\} \{n-p+a\}_{k-1} \{n-a-1\}_{n-k} (-1)^k$$
$$\times\, Q^{k(k-n+p-a-1)} R^{(k-1)(k-n-a-1)-p}$$

$$
\begin{aligned}
&= S(n, p+1, a+1) \\
&\quad - \{n\} \sum_k \left\{ {n-1 \atop k} \right\} \{n-p+a\}_k \{n-a-1\}_{n-k-1} (-1)^k \\
&\quad \times Q^{k(k-n+p-a+1)-a} R^{k(k-n-a)} Q^{p-n} R^{-p}.
\end{aligned}
$$

The two second sums are proportional by the factor $Q^{p-n}R^{-p}$, and the two first sums are both proportional to $S(n, p+1, a+1)$.

By eliminating the two second sums we obtain the formula

$$
(Q^{n-p}+R^{n-p})S(n,p+1,a+1) = R^{-p}S(n,p,a)+Q^{n-p}S(n,p,a+1). \quad \text{(B.39)}
$$

Iteration of this formula m times yields the expression

$$
\begin{aligned}
&\prod_{j=1}^{m} \left(Q^{n-p+j} + R^{n-p+j}\right) S(n,p,a) \\
&\quad = \sum_i \left\{ {m \atop i} \right\} Q^{\sum_{j=1}^{i}(n+j-p)} R^{\sum_{j=1}^{m-i}(j-p)} S(n, p-m, a-m+i). \quad \text{(B.40)}
\end{aligned}
$$

For $m = 0$ we get a trivial identity, and for $m = 1$ we get (B.39) for $p - 1$ and $a - 1$.

We prove (B.40) by induction on m. We apply (B.39) to $S(n, p-m, a-m+i)$, and obtain

$$
\begin{aligned}
&\prod_{j=1}^{m+1} \left(Q^{n-p+j} + R^{n-p+j}\right) S(n,p,a) \\
&\quad = \sum_i \left\{ {m \atop i} \right\} Q^{\sum_{j=1}^{i}(n+j-p)} R^{\sum_{j=1}^{m-i}(j-p)} R^{m+1-p} \\
&\qquad\qquad \times S(n, p-m-1, a-m-1+i) \\
&\qquad + \sum_i \left\{ {m \atop i} \right\} Q^{\sum_{j=1}^{i}(n+j-p)} R^{\sum_{j=1}^{m-i}(j-p)} Q^{n+m+1-p} \\
&\qquad\qquad \times S(n, p-m-1, a-m+i) \\
&\quad = \sum_i \left\{ {m \atop i} \right\} Q^{\sum_{j=1}^{i}(n+j-p)} R^{\sum_{j=1}^{m-i}(j-p)} R^{m+1-p} \\
&\qquad\qquad \times S(n, p-m-1, a-m-1+i) \\
&\qquad + \sum_i \left\{ {m \atop i-1} \right\} Q^{\sum_{j=1}^{i-1}(n+j-p)} R^{\sum_{j=1}^{m-i+1}(j-p)} Q^{n+m+1-p} \\
&\qquad\qquad \times S(n, p-m-1, a-m-1+i)
\end{aligned}
$$

$$= \left(\sum_i Q^{\sum_{j=1}^{i}(n+j-p)} R^{\sum_{j=1}^{m+1-i}(j-p)} S(n, p-m-1, a-m-1+i)\right)$$
$$\times \left(\left\{ m \atop i \right\} R^i + \left\{ m \atop i-1 \right\} Q^{m+1-i}\right)$$
$$= \sum_i \left\{ m+1 \atop i \right\} Q^{\sum_{j=1}^{i}(n+j-p)} R^{\sum_{j=1}^{m+1-i}(j-p)}$$
$$\times S(n, p-m-1, a-m-1+i), \tag{B.41}$$

by substitution of $i+1$ for i and using (B.18).

Now we apply (B.41) to $m = p$, and rewrite the product using (B.14) as

$$\prod_{j=1}^{p} \left(Q^{n-p+j} + R^{n-p+j}\right) = \prod_{j=0}^{p-1} \left(Q^{n-j} + R^{n-j}\right) = \frac{\{2n, 2\}_p}{\{n\}_p}.$$

Furthermore, we reverse the order of summation and apply (B.32) to $S(n, 0, a-i)$. This yields

$$S(n, p, a) = \frac{\{n\}_p}{\{2n, 2\}_p} \sum_i \left\{ p \atop i \right\} Q^{(p-i)(n-p)+\binom{p+1-i}{2}} R^{-pi+\binom{i+1}{2}}$$
$$\times \{n-a+i-1, 2\}_n \frac{\{2n, 2\}_n}{\{n\}_n}.$$

Canceling common factors proves (B.37).

(2) For $p < 0$, let

$$S(n, a, p, x) := \sum_k \left\{ n \atop k \right\} \{n+a-p\}_k \{n-a\}_{n-k} x^k (QR)^{k(k-n-a)} Q^{kp}.$$

We split the sum $S(n+1, a, p+1, x)$ in two by (B.7) applied to

$$\{n+1-a\}_{n-k+1} = \{n-a\}_{n-k+1} Q^{n-k+1} + \{n-k+1\}\{n-a\}_{n-k} R^{k-a}.$$

Then we get

$$S(n+1, a, p+1, x)$$
$$= \sum_k \left\{ n+1 \atop k \right\} \{n-p+a\}_k \{n-a\}_{n-k+1} (xR/Q)^k$$
$$\times (QR)^{k(k-n-a-2)} Q^{k(p+2)+n+1}$$
$$+ \{n+1\} \sum_k \left\{ n \atop k \right\} \{n-p+a\}_k \{n-a\}_{n-k} x^k (QR)^{k(k-n-a)} Q^{kp} R^{-a}$$
$$= S(n+1, a+1, p+2, xR/Q) Q^{n+1} + S(n, a, p, x)\{n+1\} R^{-a}.$$

And we split $S(n+1, a+1, p+1, x)$ by applying (B.7) to

$$\{n-p+a+1\}_k = \{n-p+a\}_k R^k + \{k\}\{n-p+a\}_{k-1} Q^{kp}.$$

Then we get

$$\begin{aligned}
&S(n+1, a+1, p+1, x) \\
&\quad = \sum_k \left\{ {n+1 \atop k} \right\} \{n-p-a\}_k \{n-a\}_{n-k+1} x^k (QR)^{k(k-n-a-1)} Q^{kp} \\
&\qquad + \{n+1\} \sum_k \left\{ {n \atop k-1} \right\} \{n-p+a\}_{k-1} \{n-a\}_{n-k+1} x^k \\
&\qquad\quad \times (QR)^{k(k-n-a-2)} Q^{kp+n-p+a+1} \\
&\quad = S(n+1, a+1, p+2, xR/Q) \\
&\qquad + \{n+1\} \sum_k \left\{ {n \atop k} \right\} \{n-p+a\}_k \{n-a\}_{n-k} x^{k+1} \\
&\qquad\quad \times (QR)^{k(k-n-a)} Q^{kp} R^{-n-a-1} \\
&\quad = S(n+1, a+1, p+2, xR/Q) + S(n, a, p, x) x \{n+1\} R^{-n-a-1}.
\end{aligned}$$

By eliminating $S(n+1, a+1, p+2, xR/Q)$ we obtain the formula

$$\begin{aligned}
&(R^{-a}\{n+1\} - xQ^{n+1}R^{-a-n-1}\{n+1\}) S(n, a, p, x) \\
&\quad = S(n+1, a, p+1, x) - Q^{n+1} S(n+1, a+1, p+1, x),
\end{aligned}$$

which can be rewritten as

$$S(n, a, p, x) = \frac{S(n+1, a, p+1, x) - Q^{n+1} S(n+1, a+1, p+1, x)}{\{n+1\}(R^{n+1} - xQ^{n+1})} R^{a+n+1}.$$

For $x = -1$ we have $\{n+1\}(Q^{n+1} + R^{n+1}) = \{2(n+1)\}$, so we obtain

$$\begin{aligned}
&S(n, a, p, -1) \\
&\quad = \frac{S(n+1, a, p+1, -1) - Q^{n+1} S(n+1, a+1, p+1, -1)}{\{2n+2\}} R^{a+n+1}.
\end{aligned} \tag{B.42}$$

Iteration of this formula m times yields the expression

$$\begin{aligned}
S(n, a, p, -1) = {} & \frac{R^{m(a+m)+\binom{m+1}{2}}}{\{2(n+m), 2\}_m} \sum_i \left\{ {m \atop i} \right\} (-1)^i Q^{in+\binom{i+1}{2}} R^{\binom{i}{2}} \\
& \times S(n+m, a+i, p+m, -1).
\end{aligned} \tag{B.43}$$

For $m=0$, (B.43) is a trivial identity, and for $m=1$, we get (B.42).

We prove (B.43) by induction with respect to m. First, apply (B.42) to $S(n+m, a+i, p+m, -1)$ and obtain

$$S(n,a,p,-1) = \frac{R^{m(a+n)+\binom{m+1}{2}}}{\{2(n+m),2\}_m} \sum_i \left\{ {m \atop i} \right\} (-1)^i Q^{in+\binom{i+1}{2}} R^{\binom{i}{2}+a+n+m+i+1}$$
$$\times \Big(S(n+m+1, a+i, p+m+1, -1)$$
$$- Q^{n+m+1} S(n+m+1, a+i+1, p+m+1, -1) \Big) \Big/ \{2(n+m),2\}$$
$$= \frac{R^{(m+1)(a+n)+\binom{m+2}{2}}}{\{2(n+m+1),2\}_{m+1}} \sum_i S(n+m+1, a+i, p+m+1, -1)(-1)^i$$
$$\times \left(\left\{ {m \atop i} \right\} Q^{in+\binom{i+1}{2}} R^{\binom{i}{2}+i} + \left\{ {m \atop i-1} \right\} Q^{in+\binom{i}{2}+m+1} R^{\binom{i}{2}} \right)$$
$$= \frac{R^{(m+1)(a+n)+\binom{m+2}{2}}}{\{2(n+m+1),2\}_{m+1}} \sum_i S(n+m+1, a+i, p+m+1, -1)(-1)^i$$
$$\times Q^{in+\binom{i+1}{2}} R^{\binom{i}{2}} \left\{ {m+1 \atop i} \right\}.$$

Choosing $m=-p$ in (B.43) and using the formula (B.32), we obtain

$$S(n,a,p,-1)$$
$$= \frac{R^{-p(a+n)+\binom{1-p}{2}}}{\{2(n-p),2\}_{-p}} \sum_i \left\{ {-p \atop i} \right\} (-1)^i Q^{in+\binom{i+1}{2}} R^{\binom{i}{2}} S(n-p, a+i, 0, -1)$$
$$= \frac{R^{-p(a+n)+\binom{1-p}{2}}}{\{2(n-p),2\}_{-p}} \sum_i \left\{ {-p \atop i} \right\} (-1)^i Q^{in+\binom{i+1}{2}} R^{\binom{i}{2}}$$
$$\times \{n-p-a-1-i, 2\}_{n-p} \frac{\{2(n-p),2\}_{n-p}}{\{n-p\}_{n-p}}$$
$$= \frac{R^{-p(a+n)+\binom{1-p}{2}} \{2n,2\}_n}{\{n-p\}_{n-p}}$$
$$\times \sum_i \left\{ {-p \atop i} \right\} (-1)^i Q^{in+\binom{i+1}{2}} R^{\binom{i}{2}} \{n-p-a-1-i, 2\}_{n-p}.$$

This completes the proof of (B.38). □

B.9 A Basic Transformation of a II(2,2,z) Sum

Theorem B.13. *For arbitrary a, b, x, and y and nonnegative integers n and m,*

$$\sum_{k=m}^{n} \begin{Bmatrix} n-m \\ k-m \end{Bmatrix} \{a\}_{k-m}\{b\}_{n-k} x^{k-m} y^{n-k} (QR)^{-(k-m)(n-k)}$$

$$= \sum_{k=m}^{n} \begin{Bmatrix} n-m \\ k-m \end{Bmatrix} \{n-m-a-b-1\}_{k-m}\{b\}_{n-k}(-x)^{k-m}$$

$$\times\, Q^{(k-m)(a-n+m+1)+\binom{k-m}{2}} R^{(k-m)(b-n+k)-\binom{n-m}{2}+a(n-m)}$$

$$\times \prod_{j=0}^{n-k-1} (yR^{j-a} - xQ^{j-b}). \tag{B.44}$$

Also,

$$\sum_{k=m}^{n} \begin{Bmatrix} n-m \\ k-m \end{Bmatrix} \{a\}_{k-m}\{b\}_{n-k} x^{k-m} y^{n-k} (QR)^{-(k-m)(n-k)}$$

$$= \sum_{k=m}^{n} \begin{Bmatrix} n-m \\ k-m \end{Bmatrix} \{a\}_{k-m}\{n-m-a-b-1\}_{n-k}(-y)^{n-k}$$

$$\times\, Q^{(n-k)(b-n+m+1)+\binom{n-k}{2}} R^{(n-k)(a-k+m)-\binom{n-m}{2}+b(n-m)}$$

$$\times \prod_{j=0}^{k-m-1} (xR^{j-b} - yQ^{j-a}), \tag{B.45}$$

or, equivalently,

$$\sum_{k=m}^{n} \begin{Bmatrix} n-m \\ k-m \end{Bmatrix} \{a\}_{k-m}\{b\}_{n-k} x^{k-m} y^{n-k}$$

$$\times\, (QR)^{-(k-m)(n-k)} Q^{b(k-m)} R^{a(n-k)}$$

$$= \sum_{k=m}^{n} \begin{Bmatrix} n-m \\ k-m \end{Bmatrix} \{n-m-a-b-1\}_{k-m}\{b\}_{n-k}(-x)^{k-m}$$

$$\times\, (QR)^{-(k-m)(n-k-b)} Q^{a(k-m)-\binom{k-m}{2}} R^{a(n-m)-\binom{n-m}{2}}$$

$$\times \prod_{j=0}^{n-k-1} (yR^{j} - xQ^{j}). \tag{B.46}$$

The form (B.44) is symmetric in Q and R, and reversal of the direction of summation corresponds to the exchange of a with b and x with y.

Proof: It is sufficient to prove (B.44) for $m = 0$. In the left-hand expression we substitute (B.9) in the form $\{a\}_k = (-1)^k\{-a+k-1\}_k(QR)^{ka-\binom{k}{2}}$. Next we apply the Chu–Vandermonde convolution (8.1) to the factorial and get

$$\{-a+k-1\}_k = \sum_j \begin{Bmatrix} k \\ j \end{Bmatrix} \{n-a-b-1\}_j\{b-n+k\}_{k-j} \\ \times Q^{(n-a-b-1-j)(k-j)}R^{(b-n+j)j}.$$

Substitution in the left-hand expression yields, after exchanging the order of summation,

$$\sum_j \{n-a-b-1\}_j(-x)^j \sum_k \begin{Bmatrix} n \\ k \end{Bmatrix}\begin{Bmatrix} k \\ j \end{Bmatrix}\{b\}_{n-k}\{b-n+k\}_{k-j}(-x)^{k-j}y^{n-k} \\ \times Q^{-k(n-k)+ka-\binom{k}{2}+(n-a-b-1-j)(k-j)}R^{-k(n-k)+ka-\binom{k}{2}+(b-n+j)j}.$$

Now we apply (B.19) to

$$\begin{Bmatrix} n \\ k \end{Bmatrix}\begin{Bmatrix} k \\ j \end{Bmatrix} = \begin{Bmatrix} n \\ j \end{Bmatrix}\begin{Bmatrix} n-j \\ k-j \end{Bmatrix},$$

and (B.12) to

$$\{b\}_{n-k}\{b-n+k\}_{k-j} = \{b\}_{n-j},$$

to obtain the expression

$$\sum_j \begin{Bmatrix} n \\ j \end{Bmatrix}\{n-a-b-1\}_j\{b\}_{n-j}Q^{(a-n+1)j+\binom{j}{2}}R^{(b-n+j)j-\binom{n}{2}+an} \\ \times (-x)^j \sum_k \begin{Bmatrix} n-j \\ k-j \end{Bmatrix}(-x)^{k-j}y^{n-k}Q^{\binom{k-j}{2}-b(k-j)}R^{\binom{n-k}{2}-a(n-k)}.$$

Now we apply the basic binomial theorem (B.22) to obtain (B.44).

Formula (B.45) follows from (B.44) by reversing the order of summation, exchanging a with b and x with y. Substitution of $x := xQ^b/R^a$ in (B.44) and multiplication by $R^{a(n-m)}$ gives (B.46). $\square$

B.10 A Basic Gauss Theorem

Theorem B.14. *For arbitrary a and integers p, m, n with $0 \le m \le n$,*

$$\begin{aligned}
&\sum_{k=m}^{n} \left\{ {n-m \atop k-m} \right\} \{2a, 2\}_{k-m} \{n-m-p-2a-1\}_{n-k} \\
&\quad \times (QR)^{k^2+(n-2k)(a+m)+np-\binom{n}{2}} Q^{-k(p+1)+ma+\binom{m+1}{2}} R^{-nk+\binom{k}{2}+n(a+m)-pm} \\
&= (-\sigma(p))^{n-m} \sum_{j=\lceil \frac{n-m-|p|}{2} \rceil}^{\lfloor \frac{n-m}{2} \rfloor} \left\{ {|p| \atop n-m-2j} \right\} \{n-m\}_{n-m-2j} \\
&\quad \times \{2(a+(p\wedge 0)), 2\}_j \{-1, 2\}_j \\
&\quad \times Q^{(a+(p\wedge 0)-2j)(n-m-2j)} R^{(|p|-n+m+2j)2j+(n-m)(p\wedge 0)}.
\end{aligned}$$

Proof: We substitute $b := a + p$ in the right side of (B.45). Substitution of $x := -R^p$ and $y := 1$ gives the left side of (B.31), so we can apply the quasi-symmetric Kummer formula. Eventually we apply (B.14) to the product. □

B.11 A Basic Bailey Theorem

Theorem B.15. *For arbitrary a and integers p, m, n with $0 \le m \le n$,*

$$\begin{aligned}
&\sum_{k=m}^{n} \frac{\{2(n-m), 2\}_{k-m}}{\{k-m\}_{k-m}} \{a\}_{k-m} \{m-n+p-1\}_{n-k} \\
&\times (QR)^{(n-m)(n-a-k)-n(p+k)+\binom{k-m}{2}+\binom{n+1}{2}} Q^{kp+\binom{k}{2}} R^{k(k-m)+a(n-k)+mp+\binom{m}{2}} \\
&= \begin{cases}
(-1)^{n-m} \dfrac{\{2(n-m-p), 2\}_{n-m-p}}{\{n-m-p\}_{n-m-p}} \\
\quad \times \displaystyle\sum_{j=0}^{p} \left\{ {p \atop j} \right\} \{2(n-m)-p-a-1+j, 2\}_{n-m} \\
\quad \times Q^{(p-j)(n-m-p)+\binom{p+1-j}{2}} R^{-pj+\binom{j+1}{2}} & \text{for } p \ge 0, \\
(-1)^{n-m} \dfrac{\{2(n-m), 2\}_{n-m}}{\{n-m-p\}_{n-m-p}} R^{-p(a+p)+\binom{1-p}{2}} \\
\quad \times \displaystyle\sum_{j=0}^{-p} \left\{ {-p \atop j} \right\} (-1)^j \{2(n-m-p)-a-p-1-j, 2\}_{n-m-p} \\
\quad \times Q^{j(n-m)+\binom{j+1}{2}} R^{\binom{j}{2}} & \text{for } p < 0.
\end{cases}
\end{aligned}$$

Proof: We substitute $b := 2(n-m)-a-p$ in the right side of (B.45) and choose $x := -(QR)^{n-m-a}R^{-p}$ and $y := 1$. Then apply (B.14) on the product, and apply the quasi-balanced Kummer identity (B.37), (B.38) with $a := a+p-n+m$. □

B.12 Notes

In the ordinary (non-basic) case, the binomial theorem and the Chu–Vandermonde convolution are very well-known results [Vandermonde 72] (cf. Theorem 8.1). The Chu–Vandermonde convolution with arbitrary step length is also a generalization of the binomial theorem. We have not seen this nor the generalization to commutative rings before.

The symmetric Kummer identity and the balanced Kummer identity are special cases of Kummer's theorem for Gaussian hypergeometric series, Theorem 8.7 [Vandermonde 72]:

$$\begin{aligned}{}_2F_1[a,b;1+a-b;-1] &= \sum_{n=0}^{\infty}\frac{(a)_n(b)_n(-1)^n}{n!(1+a-b)_n}\\ &= 2^{-a}\frac{\Gamma\left(\frac{1}{2}\right)\Gamma(1+a-b)}{\Gamma\left(\frac{1+a}{2}\right)\Gamma\left(1+\frac{a}{2}-b\right)}\\ &= \frac{\Gamma(1+a-b)\Gamma\left(1+\frac{a}{2}\right)}{\Gamma(1+a)\Gamma\left(1+\frac{a}{2}-b\right)}.\end{aligned}$$

The first of these two evaluations of ${}_2F_1[a,b;1+a-b;-1]$ yields the symmetric Kummer identity if we substitute $-n$ for a. The second evaluation yields the balanced Kummer identity if we substitute $-n$ for b.

We believe that the quasi-symmetric and the quasi-balanced Kummer identities are new, although special cases with numerically small values of the parameter p can be found for example in Gould's table of Combinatorial Identities [Gould 72b].

Special cases of the transformation formula in Section 12.2 can be found in Gould [Gould 72b]. However, we have not seen the general form before.

Using this transformation formula we have the Gauss identity and the Bailey identity almost as corollaries of the quasi-symmetric and the quasi-balanced Kummer identity.

When the parameter p vanishes, the Gauss identity is a special case of Gauss' evaluation of

$${}_2F_1\left[a,b;\tfrac{1+a+b}{2};\tfrac{1}{2}\right] = \frac{\Gamma\left(\frac{1}{2}\right)\Gamma\left(\frac{1+a+b}{2}\right)}{\Gamma\left(\frac{1+a}{2}\right)\Gamma\left(\frac{1+b}{2}\right)},$$

and Bailey's identity is a special case of Bailey's evaluation of

$$ {}_2F_1[a, 1-a; b; \tfrac{1}{2}] = \frac{\Gamma(\frac{b}{2})\Gamma(\frac{1+b}{2})}{\Gamma(\frac{a+b}{2})\Gamma(\frac{1-a+b}{2})}. $$

The basic binomial theorem is due to Cauchy [Cauchy 43] and Heine [Heine 47]. Heine also proved the basic version of Gauss' summation formula for hypergeometric series [Heine 47]. The basic version of Kummer's theorem is due to Bailey [Bailey 41] and Daum [Daum 42]. The basic versions of Gauss' second theorem and Bailey's theorem are due to G. Andrews [Andrews 74].

Further references can be found in [Gasper and Rahman 90]. In 1921, F. H. Jackson proved a basic version of Dougall's theorem [Jackson 21], (see also [Slater 66, p. 94]).

Bibliography

[Abel 39] Niels Henrik Abel. *Démonstration d'une expression de laquelle la formule binome est un cas particulier, in: Œuvres complètes Rédigées par ordre du roi, par B. Holmboe, Tome Premier.* Christiania: Chr. Gröndahl, 1839.

[Abramowitz and Stegun 65] Milton Abramowitz and Irene A. Stegun. *Handbook of Mathematical Functions.* New York: Dover Publications, Inc., 1965.

[Al-Salam 57] W. A. Al-Salam. "Note on a q-identity." *Math. Scand.* 5 (1957), 202–204.

[Alkan 95] Emre Alkan. "Problem no. 10473." *Amer. Math. Monthly* 102 (1995), 745.

[Andersen and Larsen 93] Erik Sparre Andersen and Mogens Esrom Larsen. "A finite sum of products of binomial coefficients, Problem 92-18, by C. C. Grosjean, Solution." *SIAM Rev.* 35 (1993), 645–646.

[Andersen and Larsen 94] Erik Sparre Andersen and Mogens Esrom Larsen. "Rothe–Abel–Jensen identiteter." *Normat* 42 (1994), 116–128.

[Andersen and Larsen 95] Erik Sparre Andersen and Mogens Esrom Larsen. "Problem no. 10466." *Amer. Math. Monthly* 102 (1995), 654.

[Andersen and Larsen 97] Erik Sparre Andersen and Mogens Esrom Larsen. "A surprisingly simple summation solution, Problem no. 10424." *Amer. Math. Monthly* 104 (1997), 466–467.

[Andersen 89] Erik Sparre Andersen. "Classification of combinatorial identities." Technical Report 26, Copehagen University Mathematics Institute, 1989. KUMI Preprint Series.

[Andrews 74] George E. Andrews. "Applications of basic hypergeometric functions." *SIAM Rev.* 16 (1974), 442–484.

[Bailey 35] W. N. Bailey. *Generalized Hypergeometric Series.* Cambridge: Cambridge University Press, 1935.

[Bailey 41] W. N. Bailey. "A note on certain q-identities." *Quart. J. Math.* 12 (1941), 173–175.

[Bang 95] Bang Seung-Jin. "Problem no. 10490." *Amer. Math. Monthly* 102 (1995), 930.

[Bernoulli 13] Jacobi Bernoulli. *Ars Conjectandi.* Baseleæ: Universitätsbibliothek Basel, 1713.

[Cauchy 43] Augustin-Louis Cauchy. "Mémoire sur les fonctions dont plusieurs valeurs sont liées entre elles par une équation linéaire, et sur diverses transformations de produits composés d'un nombre indéfini de facteurs." *C. R. Acad. Sci. Paris* 17 (1843), 523–531.

[Cauchy 87] Augustin-Louis Cauchy. *Application du calcul des résidus a la sommation de plusieurs suites, dans: Œuvres complètes IIE Série Tome VI.* Paris: Gauthier-Villars, 1887.

[Chen 94] Chen Kwang-Wu. "Problem no. 10416." *Amer. Math. Monthly* 101, 9, November (1994), 912.

[Chu 03] Chu Shih-Chieh. "Ssu Yuan Yü Chien (Precious Mirror of the Four Elements)." China, 1303.

[Daum 42] J. A. Daum. "The basic analogue of Kummer's theorem." *Bull. Amer. Math. Soc.* 48 (1942), 711–713.

[Davis 62] H. T. Davis. *The Summation of Series.* San Antonio, TX: Principia Press, 1962.

[Dixon 03] A. C. Dixon. "Summation of a certain series." *Proc. London Math. Soc.* 35 (1903), 285–289.

[Doster 94] David Doster. "Problem no. 10403." *Amer. Math. Monthly* 101 (1994), 792.

[Dougall 07] J. Dougall. "On Vandermonde's theorem and some more general expansions." *Proc. Edin. Math. Soc.* 25 (1907), 114–132.

[Edwards 87] A. W. F. Edwards. *Pascal's Arithmetical Triangle.* London: Charles Griffin & Co., Ltd., 1987.

[Ekhad 94] Shalosh B. Ekhad. "Problem no. 10356." *Amer. Math. Monthly* 101 (1994), 75.

[Fjeldstad 48] Jonas Ekman Fjeldstad. "Løste Oppgaver 4." *Norsk Mat. Tidsskr.* 30 (1948), 94–95.

[Galperin and Gauchman 04] Gregory Galperin and Hillel Gauchman. "Problem no. 11103." *Amer. Math. Monthly* 111 (2004), 725.

[Gasper and Rahman 90] George Gasper and Mizan Rahman. *Basic Hypergeometric Series, Encyclopedia of Mathematics and its Applications, vol. 35.* Cambridge: Cambridge University Press, 1990.

[Gauss13] C. F. Gauss. "Disquisitiones generales circa seriem infinitam $1+\frac{\alpha\beta}{1\cdot\gamma}x+\frac{\alpha(\alpha+1)\beta(\beta+1)}{1\cdot 2\cdot\gamma(\gamma+1)}xx+\frac{\alpha(\alpha+1)(\alpha+2)\beta(\beta+1)(\beta+2)}{1\cdot 2\cdot 3\cdot\gamma(\gamma+1)(\gamma+2)}x^3+\cdots$." *Comm. soc. reg. sci. Gött. rec.* II (1813), 123–162.

[Gessel and Stanton 82] Ira M. Gessel and Dennis Stanton. "Strange evaluations of hypergeometric series." *SIAM J. Math. Anal.* 13 (1982), 295–308.

[Gessel 95a] Ira M. Gessel. "Finding identities with the WZ method." *J. Symbolic Comput.* 20 (1995), 537–566.

[Gessel 95b] Ira M. Gessel. "Problem no. 10424." *Amer. Math. Monthly* 102 (1995), 70.

[Golombek 94] Renate Golombek. "Problem no. 1088." *Elem. Math.* 49 (1994), 126–127.

[Gosper 78] R. William Gosper, Jr. "Decision procedure for indefinite hypergeometric summations." *Proc. Nat. Acad. Sci. U.S.A.* 75 (1978), 40–42.

[Gould 72a] Henry W. Gould. "The case of the strange binomial identities of Professor Moriarty." *Fibonacci Quart.* 10 (1972), 381–391, 402.

[Gould 72b] Henry W. Gould. *Combinatorial Identities.* Morgantown, WV: Gould Publications, 1972.

[Gould 74] Henry W. Gould. "The design of the four binomial identities: Moriarty intervenes." *Fibonacci Quart.* 12 (1974), 300–308.

[Gould 76] Henry W. Gould. "Sherlock Holmes and the godfather of organised crime." *The Baker Street Journal* 26 (1976), 34–36.

[Graham et al. 94] Ronald Lewis Graham, Donald Knuth, and Oren Patashnik. *Concrete Mathematics*, Second edition. Reading, MA: Addison-Wesley Publishing Company, 1994.

[Grosjean 92] C. C. Grosjean. "Problem no. 92–18." *SIAM Rev.* 34 (1992), 649.

[Hagen 91] John George Hagen. *Synopsis der Höheren Mathematik, Bd. I.* Berlin: Verlag von Felix L. Dames, 1891.

[Hardy 23] Godfrey Harold Hardy. "A chapter from Ramanujan's note-book." *Proc. Cambridge Phil. Soc.* 21 (1923), 492–503.

[Heine 47] E. Heine. "Untersuchungen über die Reihe $1 + \frac{(1-q^\alpha)(1-q^\beta)}{(1-q)(1-q^\gamma)} \cdot x + \frac{(1-q^\alpha)(1-q^{\alpha+1})(1-q^\beta)(1-q^{\beta+1})}{(1-q)(1-q^2)(1-q^\gamma)(1-q^{\gamma+1})} \cdot x^2 + \cdots$." *J. Reine Angew. Math.* 34 (1847), 285–328.

[Heine 78] E. Heine. *Theorie der Kugelfunctionen und der verwandten Functionen.* Berlin: G. Reimer, 1878.

[Hoe 77] John Hoe. *Les Systèmes d'équations polynômes dans le Siyuan Yujian (1303).* Paris: Collège de France, Institut des hautes Études Chinoises, 1977.

[Jackson 21] F. H. Jackson. "Summation of q-hypergeometric series." *Mess. Math.* 50 (1921), 101–112.

[Jensen 02] Johan Ludvig William Valdemar Jensen. "Sur une identité d'Abel et sur d'autres formules analogues." *Acta Math.* 26 (1902), 307–318.

[Kaucký 75] Josef Kaucký. *Kombinatorické identity.* Bratislava: Veda, Vydavatel'stvo Slovenskej Akadémie Vied, 1975.

[Knuth 93] Donald E. Knuth. "Johann Faulhaber and sums of powers." *Math. of Computation* 61 (1993), 277–294.

[Koornwinder 91] Tom H. Koornwinder. *Handling hypergeometric series in Maple*, IMACS Ann. Comput. Appl. Math., 9, pp. 73–80. Basel: Baltzer, 1991.

[Krall 38] H. L. Krall. "Differential equations for Chebyshef polynomials." *Duke Math. J.* 4 (1938), 705–718.

[Kummer 36] E. E. Kummer. "Über die hypergeometrische Reihe $F(\alpha, \beta, \gamma, x)$." *J. Reine u. Angew Math.* 15 (1836), 39–83, 127–172.

[Kvamsdal 48] Johannes Kvamsdal. "Løste Oppgaver 4." *Norsk Mat. Tidsskr.* 30 (1948), 113–114.

[Larcombe and Larsen 07] Peter J. Larcombe and Mogens E. Larsen. "Some binomial coefficient identities of specific and general type." *Util. Math.* 74 (2007), To appear.

[Larcombe et al. 05] Peter J. Larcombe, Mogens E. Larsen, and Eric J. Fennessey. "On two classes of identities involving harmonic numbers." *Util. Math.* 67 (2005), 65–80.

[Lavoie 87] J. L. Lavoie. "Some summation formulas for the series ${}_3F_2(1)$." *Math. Comp.* 49 (1987), 269–274.

[Ljunggren 47] Wilhelm Ljunggren. "Oppgave 4." *Norsk Mat. Tidsskr.* 29 (1947), 122.

[Ma and Wang 95] Ma Xin-Rong and Wang Tian-Ming. "Problem no. 95–1." *SIAM Rev.* 37:1, 98.

[Petkovšek et al. 96] Marco Petkovšek, Herbert S. Wilf, and Doron Zeilberger. $A = B$. Wellesley, MA: A K Peters, 1996.

[Pfaff 97] J. F. Pfaff. "Observationes analyticae ad L. Euleri Institutiones Calculi Integralis." *Nova acta acad. sci. Petropolitanae* IV supplem. II. et IV. (1797), 38–57.

[Riordan 68] John Riordan. *Combinatorial Identities.* New York: John Wiley & Sons, Inc., 1968.

[Rothe 93] Heinrich August Rothe. *Formulae de serierum reversione demonstratio universalis signis localibus combinatorio-analyticorum vicariis exhibita.* Leipzig: Litteraturis Sommeriis, 1793.

[Saalschütz 90] L. Saalschütz. "Eine Summationsformel." *Zeitschr. Math. Phys.* 35 (1890), 186–188.

[Santmyer 94] Joseph M. Santmyer. "Problem no. 10363." *Amer. Math. Monthly* 101:2 (1994), 175.

[Sinyor and Speevak 01] Joseph Sinyor and Ted Speevak. "A new combinatorial identity." *Int. J. Math. Math. Sci.* 25 (2001), 361–363.

[Slater 66] Lucy Joan Slater. *Generalized Hypergeometric Functions.* Cambridge: Cambridge University Press, 1966.

[Stanley 97] Richard P. Stanley. *Enumerative Combinatorics, Volume 1.* New York: Cambridge University Press, 1997.

[Staver 47] Tor B. Staver. "Om summasjon av potenser av binomialkoeffisientene." *Norsk Mat. Tidsskr.* 29 (1947), 97–103.

[Vandermonde 72] A. T. Vandermonde. "Mémoire sur des irrationnelles de différens ordres avec une application au cercle." *Mèm. Acad. Roy. Sci. Paris* (1772), 489–498.

[Wachs and White 91] Michelle Wachs and Dennis White. "p, q-Stirling numbers and set partition statistics." *J. Combin. Theory Ser. A* 56 (1991), 27–46.

[Watson 24] G. N. Watson. "Dixon's theorem on generalized hypergeometric functions." *Proc. London Math. Soc. (2)* 23 (1924), 31–33.

[Watson 25] G. N. Watson. "A note on generalized hypergeometric series." *Proc. London Math. Soc. (2)* 23 (1925), 13–15.

[Whipple 25] F. J. W. Whipple. "A group of generalized hypergeometric series; relations between 120 allied series of the type $F(a, b, c; e, f)$." *Proc. London Math. Soc. (2)* 23 (1925), 104–114.

[Wilf and Zeilberger 90] Herbert S. Wilf and Doron Zeilberger. "Rational functions certify combinatorial identities." *J. Amer. Math. Soc.* 3 (1990), 147–158.

[Zeilberger 90] Doron Zeilberger. "A holonomic systems approach to special function identities." *J. Comput. Appl. Math.* 32 (1990), 321–368.

[Zeilberger 91] Doron Zeilberger. "The method of creative telescoping." *J. Symbol. Comput.* 11 (1991), 195–204.

Index